ACPL ITEM

DISCARDED   Y0-AAD-734

12-20-74

Fundamentals of Aircraft Flight

Hayden Series in Aeronautical Technology

Frederick K. Teichmann, Consulting Editor

Assistant Dean and Professor of Aeronautics and Astronautics
Undergraduate and Graduate Divisions
School of Engineering and Science
New York University

JET-ENGINE FUNDAMENTALS
Norman E. Borden, Jr.

FUNDAMENTALS OF AIRCRAFT ENVIRONMENTAL CONTROL
Alvin Ebeling

FUNDAMENTALS OF AIRCRAFT STRUCTURAL ANALYSIS
Frederick K. Teichmann

FUNDAMENTALS OF AIRCRAFT FLIGHT AND ENGINE INSTRUMENTS
Jack Andresen

FUNDAMENTALS OF AIRCRAFT PISTON ENGINES
Norman E. Borden, Jr. and Walter J. Cake

FUNDAMENTALS OF AIRCRAFT FLIGHT
Frederick K. Teichmann

Fundamentals of Aircraft Flight

FREDERICK K. TEICHMANN
New York University

HAYDEN BOOK COMPANY, INC.
Rochelle Park, New Jersey

ISBN 0-8104-5587-0 (soft-bound edition)
   0-8104-5588-9 (hard-bound edition)
Library of Congress Catalog Card Number 73-11925

Copyright © 1974 by HAYDEN BOOK COMPANY, INC. All rights reserved. No part of this book may be reprinted, or reproduced, or utilized in any form or by any electronic, mechanical, or other means, now known or hereafter invented, including photocopying and recording, or in any information storage and retrieval system, without permission in writing from the Publisher.

Printed in the United States of America

1	2	3	4	5	6	7	8	9	PRINTING
74	75	76	77	78	79	80	81	82	YEAR

Preface

While this book is intended for those who have had introductory courses in trigonometry and algebra, with perhaps some exposure to geometry and physics, none of the mathematics is particularly difficult.

Much of the material has been used by the author in college introductory courses, but anyone who wishes to study advanced courses in aerodynamics should find this material useful. It presents an overview of normal airplane flight, and thus can serve as an introduction to, and a framework of reference for, more detailed and mathematical study.

The book is written for high school and technical institute students, but it should present no difficulty to those interested in self-study. It can also be used as a ready reference book.

FREDERICK K. TEICHMANN

Contents

The Airplane 1
   Wing — Tail Surfaces — Fuselage — Landing Gear — The Power Plant — Review Questions

Properties of Air 3
   Composition of Air — Air Density — Temperature — Pressures — Effect of Pressure and Temperature — Viscosity — Modulus of Elasticity — Standard Atmosphere — Review Questions

Airflow 9
   Equation of Continuity—Streamlines—Stream Tubes—Bernouilli's Theorem—Flow around an Airfoil—Laminar Flow—Boundary Layer—Turbulence—Supersonic Flow—Mach Number—Review Questions

Motion 18
   Laws of Motion — Wind, Motion, Velocity, and Speed — Speed Regimes — Review Questions

Airfoil Geometry 21
   Chord Line — Camber — Leading Edge Radius — Trailing Edge — Thickness — Airfoil Coordinates — Review Questions

Wing Geometry 28
   Span — Root — Chord — Mean Geometric Chord — Problems — Planforms — Angles — Sweepback — Dihedral — Angle of Incidence — Wing Area — Aspect Ratio — Review Questions

Reference Axes 41
   Body Axes — Wind Axes — Review Questions

Basic Airfoil Aerodynamics 43

Angle of Attack — Aerodynamic Forces — Resultant Force — Lift and Drag Forces — Lift and Drag Equations — Presenting Aerodynamic Data — Graphical Representation — Review Questions

Advanced Airfoil Aerodynamics 54

The Lift Coefficient — Slope of the Lift Curve — Angle of Attack Variation — The Drag Coefficient — The Induced Drag — Aspect Ratio Corrections — Aerodynamic Twist — Lift-Over-Drag Ratio — Center of Pressure — Moment Coefficients — Aerodynamic Center — Section Characteristics — Recapitulation — Review Questions

Non-Airfoil Drag 70

Parasite Drag — Interference Drag — Review Questions

Lift Increase Devices 73

For Minimum Speeds — Median Camber Change — Types of Flaps — Maximum Lift Coefficient for Flaps — Review Questions

Experimental Methods 78

The Wind Tunnel—The Measuring System—Method of Operation—Models—Reynolds Number—Aerodynamic Similarity—Review Questions

Motions of the Airplane 84

Linear Motion — Curvilinear Motion — Angles — Review Questions

Rectilinear Flight 86

Vertical Forces in Horizontal Flight — Some Conclusions — Variation of Speed with Altitude — Horizontal Forces in Horizontal Flight — Variation of Thrust —Power Required — Variation of Power with Altitude — Power Coefficient — Review Questions

Encountering a Gust 102

Effect of a Gust — Change in Lift Coefficient — Effect of Gust on Lift — Load Factor — Review Questions

Climbing Flight 108

Forces Acting on the Airplane — Summation of Forces — Speed along Climb Path — Angle of Climb — Rate of Climb — Power Required — Review Questions

Gliding Flight 118

Summation of Forces — Interpreting the Equations — Effect of Thrust on the Glide — Sinking Speed — The Glide Angle — Landing — Review Questions

The Dive 127

Summation of Forces — Limiting Speed — Pulling out of a Dive — Review Questions

Curvilinear (Turning) Flight 131

Centrifugal Force — Forces on the Airplane — Summation of Forces — Analysis of Equations — Radius of Turn — Review Questions

Controls 140

The Flap — Longitudinal Control — Lateral Control — Ailerons — Variation with Angle of Attack — Adverse Yawing Moments — Cockpit Controls — Directional Controls — Hinge Moments — Servo Control Tab — Review Questions

Static Stability 151

Equilibrium — Static Stability — Static Longitudinal Stability — Pitching Moment Calculations — Reference Dimensions — Forces Involved — Moments Involved — Angles — Speeds Involved — Pitching Moment Equation — Tabular Calculations — Analysis — Degree of Stability — Effect of Thrust — Neutral Stability — Review Questions

Dynamic Stability 166

Definition — Unstable Condition — Neutral Dynamic Stability — Positive Dynamic Stability — Review Questions

Lateral and Directional Stability 170

Rolling Moments — Static Lateral Stability — Dynamic Lateral Stability — Use of Dihedral — Yawing Moments — Use of Sweepback — Review Questions

Index 173

Fundamentals
of
Aircraft Flight

THE AIRPLANE

Most readers of this book will be familiar with the common terms used in describing an airplane. A few may not be, and others may not be sure on all points. (It will be hard indeed to find a student who does not know what a wing is; not so difficult to discover one who is confused about the empennage.) For this reason, and to define the terms used in the book, definitions of the main elements of an airplane are given.

Wing

This is the main lifting element of the airplane (Fig. 1). It carries the entire airplane and its contents. Although most people think of airplanes as having two wings, a right and a left, the wing is considered a single aerodynamic element. The outer portions of the trailing edge of the wing —the ailerons—are movable and are used for control. The flaps of the trailing edge of the wing may also be movable and act as lift-increase devices.

Tail Surfaces

These elements stabilize and control the airplane. The vertical tail surfaces have a stationary or fixed forward portion, the fin, and a movable rearward portion, the rudder. The horizontal tail surfaces also have a stationary or fixed surface, the stabilizer, and a movable rearward portion, the elevator.

Together, the vertical and horizontal tail surfaces are known as the *empennage*.

Fuselage

The main body of the aircraft, housing the crew, cargo, passengers, and the like, is known as the *fuselage*.

Landing Gear

For land-based aircraft, the landing gear consists of the wheels, tires, brakes, shock absorbers, axles, supporting struts, cowlings, and retract-

Landing Gear (cont'd)

ing mechanisms. For water-based aircraft, all the elements of the landing gear are replaced by floats, sometimes called pontoons. For a flying boat, the water-supporting element is incorporated in the fuselage.

The Power Plant

This is the thrust-producing unit (the engine and propeller, or the various types of jet engines), the fuel and lubrication systems, the engine controls, etc. The power-producing unit is often housed in separate pods or *nacelles*.

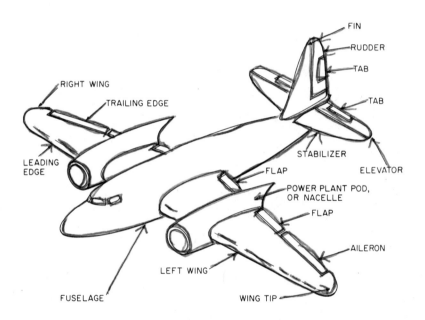

Fig. 1 External Components of an Airplane

Review Questions

1. What are the main elements of an airplane?
2. List the control units of the airplane.
3. What are the fuselage, empennage, engine nacelle, aileron?

PROPERTIES OF AIR

Airplane flight depends primarily on how much air and the manner in which it flows past a plane's wing and control surfaces. Properties of air, such as temperature, pressure, density, etc., affect these conditions. What these properties are and how they are measured or computed are described in this section. How they affect airflow is covered in the following chapter.

Composition of Air

Dry air consists of about 78 percent nitrogen, 21 percent oxygen, and 1 percent carbon dioxide and rare gases by volume. The mixture does not change with altitude, but the weight of the air does.

Air Density

Air density may be expressed in terms of its weight (w) per volume (*weight density*), or its weight-to-gravity (g = 32.2 ft per sec^2) ratio, w/g (*mass density*).

The symbol ρ (Greek rho) is used to designate *mass density*. The standard sea level *weight density* of air is 0.0764 lb per cu ft. Its mass density at sea level is

$$\rho = \frac{w}{g} = \frac{0.0764}{32.2} = 0.002377$$

Temperature

In the American system, temperatures are given in degrees Fahrenheit. Most other countries use the Celsius or Centigrade scale. To convert from one to the other use:

$$C = \frac{5}{9}(F - 32)$$

where C = degrees Celsius or Centigrade
F = degrees Fahrenheit

Celsius and Fahrenheit are the names of the inventors of the two scales

Temperature (cont'd)

based upon the freezing points (0° for Celsius and 32° for Fahrenheit) and boiling points (100° for Celsius and 212° for Fahrenheit) of water.

Absolute temperatures are based upon the theoretically lowest temperature obtainable. The physicist Rankine established this temperature to be $-459.69°$ F. Hence, to convert Fahrenheit to Rankine, add 459.69 to all Fahrenheit readings. The physicist Kelvin established the lowest temperature to be $-273.16°$, using the Celsius scale. Hence, add 273.16 to all Celsius readings to convert to the Kelvin scale. (The values 460 and 273 are used in practical calculations.)

Example: The temperature reading for a summer day is 72° F. (a) What would be the reading on the Celsius scale? (b) on the Rankine scale? (c) on the Kelvin scale?

Solution: (a) To convert to the Celsius scale, use the formula $C = 5/9(F - 32)$. Substituting 72 for F, $C = 22.22$. (Calculations have been carried out to the second decimal place, although for practical purposes one place is enough.)
   (b) for the Rankine scale: $°R = F + 460° = 532°$
   (c) for the Kelvin scale: $°K = C + 273° = 295°$

Pressures

Barometric pressures, as measured by such instruments as barometers, are often expressed in inches of mercury. At sea level, under standard conditions of pressure and temperature (14.7 pounds per square inch, 59° F), the pressure is 29.92 inches of mercury.

Effect of Pressure and Temperature

The volume of air varies inversely with the pressure when the temperature remains constant (Boyle's Law), and varies directly with the absolute temperature when the pressure remains constant (Charles' Law). These two relationships combine to form a more useful relationship for aerodynamic purposes:

$$\frac{p_1 v_1}{T_1} = \frac{p_2 v_2}{T_2}$$

where p_1, v_1, and T_1 represent values of pressure, volume, and temperature with one set of conditions, and p_2, v_2, and T_2 represent similar values when the conditions are changed.

Effect of Pressure and Temperature (cont'd)

Since density is weight divided by volume, it follows that the density of a given volume of gas is increased by increases in pressure, and is decreased by increases in temperature. In other words, the density, or specific weight, varies directly with the pressure and inversely with the absolute temperature, or

$$\frac{w_1}{w_2} = \frac{p_1 T_2}{p_2 T_1}$$

or, in terms of mass density, w/g or ρ:

$$\frac{\rho_1}{\rho_2} = \frac{p_1 T_2}{p_2 T_1}$$

Example: Under standard air conditions, the weight of air is 0.07635 pounds per cubic foot at 59° F and 29.92 inches of mercury pressure. What is the weight at 80° F and 30.2 inches of mercury?

Solution: The applicable formula is

$$\frac{w_1}{w_2} = \frac{p_1 T_2}{p_2 T_1}$$

where w_1 = 0.07635 lb per cu ft
p_1 = 29.92 in. Hg (mercury)
T_1 = 59 + 460
w_2 = to be found
p_2 = 30.2 in. Hg
T_2 = 80 + 460

Substituting, and solving for w_2, the solution is 0.0741 lb per cu ft.

Note: The pressures used were those given in inches of mercury, rather than in pounds per square inch. Since these are ratios, there should be no error as long as the same units are used.

Another form of the Boyle-Charles law, known as the *perfect-gas equation*, is

$$PV = WRT$$

Effect of Pressure and Temperature (cont'd)

where P = absolute pressure, lb per sq ft
V = volume, cu ft
W = weight, lb
R = gas constant for air = 53.34
T = absolute temperature = 459.6 + temperature in °F.

This formula can be rearranged to give

$$\frac{W}{V} = \frac{P}{RT}$$

This formula (in lb per cu ft) is particularly useful when considering compression of heated or unheated air in a closed vessel or duct.

The variation of the air density in flight at moderate speeds and at a constant altitude is so small that the air under such conditions is said to be *incompressible*. At low speed, air behaves more or less like water or any other incompressible fluid. At or above the speed of sound, air does become compressible, and marked density changes do occur.

Whether an aircraft is designed to fly below or above the speed of sound, it still has to take off and land at speeds considerably lower than the speed of sound. Consideration of the behavior of the air—even when it is incompressible—is therefore highly pertinent to all aircraft.

Viscosity

Viscosity, another property of air, should be defined, though it may be referred to only infrequently in this volume. When one layer of air slides over another, some *viscous,* or friction-like forces result. The useful quantity is known as the coefficient of viscosity, μ (the Greek letter mu), and has the value under standard air conditions at sea level of 0.3748×10^{-6} pound-second per square foot. The viscosity of air varies with temperature changes, but is practically unaffected by pressure changes.

Since calculations involving viscosity also involve the mass density of the air, another term—kinematic viscosity—is used, and is given the designation ν (the Greek letter nu), which is expressed by the equation

$$\nu = \frac{\mu}{\rho}$$

where μ is the coefficient of viscosity, as defined above, and ρ is the mass density of the air.

Viscosity (cont'd)

Numerically then,

$$\nu = \frac{0.3748 \times 10^{-6}}{0.002378} = 0.0001576 \text{ sq ft/sec at sea level}$$

Viscosity is an important factor in the study of laminar flow, aerodynamic resistance, and boundary layers.

Modulus of Elasticity

The modulus of elasticity (E) is a ratio of stress to strain, where stress is expressed in pounds per square inch, for example, and strain as the elongation per inch. For solids, such a quantity can be measured by subjecting a specimen to stress (pressure or tension), and then measuring the elongation or compression. For air, the experimental work is more complicated.

The modulus of elasticity is not often used in aerodynamics, but is mentioned here to complete the list of air properties. Sir Isaac Newton proved that the velocity of sound through a fluid or gas varies directly with the square root of the modulus of elasticity (E) and inversely with the square root of the mass density or the velocity of sound.

Standard Atmosphere

It is important to accept a common standard for measuring or considering the properties of air. Accordingly, a "standard atmosphere" has been adopted, indicating the temperature, pressure, and density of air at various altitudes. Prepared by the International Civil Aeronautics Organization (ICAO), these values are based on the yearly averages at 40° latitude in this country. The accompanying table lists various characteristics of the accepted standard atmosphere. At sea level, the "standard" value of temperature is 59°F, or 15°C, while the pressure, in inches of mercury, is 29.92 inches, the equivalent of 14.6965 pounds per square inch. (*Temperature inversions,* however, are often possible at certain altitudes, thus making the air warmer than at lower altitudes.)

Standard Atmosphere (cont'd)

ICAO Standard Atmosphere

Altitude (feet)	°F	Density (lb-sec/ft)	Kinematic Viscosity (lb-sec/sq ft)	Speed of Sound (ft/sec)
0	59	0.002377	0.000158	1116.89
1000	55.43	0.002308	0.000161	1113.05
2000	51.87	0.002241	0.000165	1109.19
3000	48.30	0.002175	0.000169	1105.31
4000	44.74	0.002111	0.000174	1101.43
5000	41.17	0.002048	0.000178	1097.53
6000	37.60	0.001987	0.000182	1093.61
7000	34.04	0.001927	0.000187	1089.68
8000	30.47	0.001868	0.000192	1085.74
9000	26.90	0.001811	0.000197	1081.78
10,000	23.34	0.001655	0.000202	1077.81
15,000	5.51	0.001496	0.000229	1057.73
20,000	−12.32	0.001266	0.000262	1037.26
25,000	−30.15	0.001065	0.000302	1016.38
30,000	−47.98	0.000889	0.000349	995.06
35,000	−65.82	0.000737	0.000405	973.28
40,000	−69.70	0.000585	0.000506	968.47
45,000	−69.70	0.000460	0.000643	968.47
50,000	−69.70	0.000362	0.000818	968.47
55,000	−69.70	0.000285	0.001040	968.47
60,000	−69.70	0.000224	0.001323	968.47
65,000	−69.70	0.000176	0.001682	968.47

Review Questions

1. Of what is air composed?
2. How is "standard atmosphere" defined?
3. What is the difference between viscosity and kinematic viscosity?
4. Is air considered compressible or incompressible? What is the dividing line?
5. What is the density (ρ) of dry air, at 0° F and 29.85 in. Hg pressure?
6. What is the pressure, in terms of inches of mercury, at 20,000 feet in standard atmosphere?

AIRFLOW

The study of *airflow* is basic to the science of aerodynamics. Some of the fundamental concepts will be considered here.

Equation of Continuity

Sometimes called the *continuity principle,* the equation of continuity states that *the mass of fluid flowing into a system is equal to the mass of fluid flowing out.* In the pipe of Fig. 2, with cross section A_1 at station 1 and A_2 at station 2, the mass of flow is

$$\rho_1 A_1 v_1 = \rho_2 A_2 v_2$$

where ρ is the mass density of the fluid (the ratio w/g, where w is the weight in pounds of the fluid per unit volume and g is the gravitational constant) and v_1 and v_2 are the air velocities at A_1 and A_2, the cross-sectional areas at stations 1 and 2, respectively.

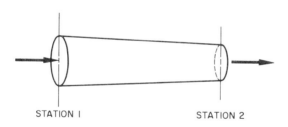

Fig. 2   Fluid Flow through a Pipe

Example: Air enters a pipe with a diameter of ten feet at a velocity of 10 miles per hour and leaves the pipe where the diameter is five feet. What is the speed of the air leaving the pipe?

Solution: Air can be considered incompressible at speeds below that of sound, so the inlet and outlet density will be the same. (The exact value is therefore unimportant to the problem.)

Equation of Continuity (cont'd)

Since $\rho_1 = \rho_2$, the continuity equation simplifies to

$$A_1 v_1 = A_2 v_2$$

$$v_2 = \frac{\pi r_1^2 v_1}{\pi r_2^2}$$

Allowing π to cancel out:

$$v_2 = (25 \times 10) / 6.25 = 40 \text{ mph}$$

Streamlines

Airflow can be thought of as consisting of a series of *streamlines* in which each of the particles moves in the same direction. In steady flow, the shape of the streamline does not change with time. If it did, the flow would be described as turbulent or unsteady. Moreover, since the particles of one streamline maintain their position relative to those of the others, streamlines do not cross.

In aerodynamics, it is often convenient to consider two-dimensional flow. If the streamline is in one plane, and the adjacent streamlines are in the same plane—or are in parallel planes and are identical in the parallel planes—the pattern of the streamlines is said to be two-dimensional.

If there are a series of streamlines, those in one plane moving faster than those in the adjacent one, the streamlines are shown closer together. This representation is used in the next section when considering airflow around an airfoil.

Stream Tubes

If there are a number of streamlines, one can arbitrarily think of a *stream tube* in a portion of the total flow, as shown in Fig. 3. If a plane perpendicular to the steady flow is taken, and a closed curve $a_1 b_1$ is considered at one station and another closed curve $a_2 b_2$ considered at a downstream station, streamlines drawn from one point on curve $a_1 b_1$ would intersect $a_2 b_2$ at corresponding points. The streamlines so drawn would form a *stream tube*. This stream tube can be seen as an imaginary pipe—a very useful concept in understanding the equation of continuity and later in considering applications of Bernouilli's Theorem.

Stream Tubes (cont'd)

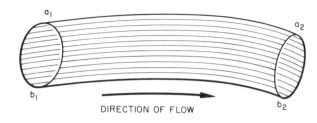

Fig. 3 A Stream Tube

The wall of the stream tube is composed of streamlines, so there can be no flow through that wall. The cross section of the tube may vary, hence—since no fluid is created, increased, nor taken away—the velocity must be greater where the cross section is small, and vice versa.

Bernouilli's Theorem

Physicist Daniel Bernouilli pointed out that when a fluid flows steadily through a pipe of varying cross section, if there is no friction, the total energy remains the same. This principle is useful in considering the airflow around any object.

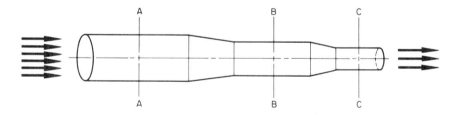

Fig. 4 Model for Bernouilli's Theorem

In Fig. 4, according to Bernouilli's theorem, the total energy at Station A-A is the same as that at station B-B and at station C-C. The total energy is made up of three forms:

1. Potential energy. This is the same as the energy due to height above a fixed reference level. For example, a ball held and then dropped would have, while held, potential energy equivalent to the height at which it was held.
2. Dynamic energy inherent in the fluid in motion.
3. Pressure energy.

Bernouilli's Theorem (cont'd)

This can be expressed mathematically in an equation:

$$(Wv^2/2g) + (Wp/w) + Wz = C_1, \text{ a constant}$$

where W represents the total weight of the fluid and v its velocity
g is the acceleration due to gravity
p is the pressure
w is the specific weight of the fluid
z is the height or altitude of the pipe section under consideration

Dividing this equation by W results in the equation:

$$(v^2/2g) + (p/w) + z = C_2, \text{ another constant}$$

Each term of this equation may be multiplied by w without changing the value of the equation:

$$(wv^2/2g) + p + wz = C_3, \text{ still another constant}$$

For horizontal flow, wz would not change. The value of w/g is the mass *density* of the fluid (designated by the letter ρ). Rearranging terms, Bernouilli's equation in its most useful form becomes

$$(\rho v^2/2) + p = C_4, \text{ a constant}$$

This equation holds for air at velocities below that of the speed of sound, at which air may be regarded as incompressible.

Flow around an Airfoil

The concepts of streamlines, stream tubes, the continuity equation, and Bernouilli's theorem help us to understand the action of airflow around an obstruction, such as an airfoil.

The streamline flow around an airfoil, shown in Fig. 5, can be demonstrated by smoke streamers from a series of orifices located upstream at station A-A. They change in direction as they approach the airfoil, the change being guided by the airfoil's contour. The streamlines are spaced closer together over the top, indicating a greater velocity, while below the wing they are spaced farther apart, indicating a lesser velocity. At some distance aft of the airfoil, the streamlines become parallel again and travel at uniform speed, , though somewhat lower than the speed ahead of the airfoil at station A-A.

Flow around an Airfoil (cont'd)

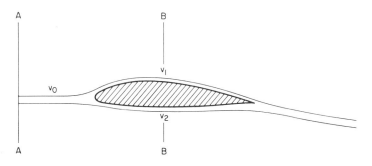

Fig. 5 Fluid Flow around an Airfoil

Designating the speed of the airflow at station A-A by v_o, and at station B-B by v_1 for the flow above the wing and v_2 for the flow below the wing, it follows that v_o is less than v_1, but greater than v_2. We can now apply Bernouilli's principle that the total energy in the stream tube in which the airfoil is located is constant at any station. Therefore, the energy at station A-A is the same as at station B-B for a stream tube encompassing the region over the top of the airfoil, and for a stream tube encompassing the region below the airfoil. For the stream tube over the airfoil,

$$(\rho v_o^2/2) + p_o = (\rho v_1^2/2) + p_1$$

from which it follows that

$$p_1 = p_o + (\rho/2)(v_o^2 - v_1^2)$$

For the stream tube below the airfoil,

$$(\rho v_o^2/2) + p_o = (\rho v_2^2/2) + p_2 \quad \text{or} \quad p_2 = p_o + (\rho/2)(v_o^2 - v_2^2)$$

The pressure in the air stream adjacent to and over the top of the airfoil is less than that of the free air stream. It is sometimes called *suction pressure*, or just plain *suction*. The pressure in the air stream below the wing adjacent to the bottom of the airfoil is greater than that in the free stream. This pressure would push upward on the airfoil.

The difference in pressures equals the total pressure on the airfoil, or

$$p_2 - p_1 = (\rho/2)(v_1^2 - v_2^2)$$

Flow around an Airfoil (cont'd)

The pressures on the airfoil or in the airstream are proportional to the square of the velocity. (It is convenient to refer pressures to the free stream velocity rather than to local velocities.)

Given enough time, pressures could be calculated, but it is far quicker and more accurate to obtain the pressures experimentally, either with a suitable model in a wind tunnel or in full-scale flight.

Laminar Flow

When moving air comes in contact with a surface, its behavior becomes important to the aerodynamicist, since it can give an insight as to why better results are obtained in one case than in another.

Suppose a thin plate is placed in a uniform flow of air. As the air flows over the plate, friction is encountered. The flow of the layer next to the plate is retarded and brought to zero due to friction of the plate. Due to the friction of the air molecules, the adjacent air is also retarded, but the effect diminishes rapidly the farther away the air is from the plate. This change in flow can be represented graphically, as in Fig. 6, where the length of the arrows represents velocity of flow.

Ahead of the plate (station 1) the air flow is uniform. Each parallel stream has the same velocity. But at station 2, the friction at the plate has slowed down parallel streams as shown. If any following station were examined, the flow would be the same as at station 2. This type of flow is said to be *laminar* (sheet-like), as opposed to turbulent. This effect is apparent over any surface, and is greater for rough surfaces than for smooth. It is reflected in an increase of aerodynamic resistance suffered by the thin plate or any other object.

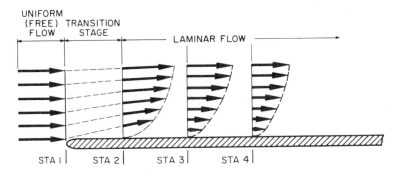

Fig. 6 Velocity Profiles—Laminar Flow along a Thin Smooth Plate

Boundary Layer

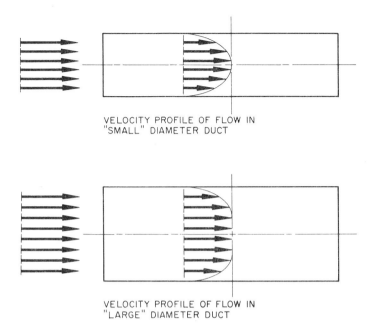

Fig. 7 Effect of Laminar Flow in Ducts

The region in which the flow near a surface is reduced is the *boundary layer*. It is desirable to reduce the thickness of this layer by reducing the surface friction, or by sucking away the boundary layer.

If a pipe or duct is considered, the thickness of the boundary layer may tend to choke or impede the flow (as compared to the velocity of the free stream airflow) if the diameter of the pipe or duct is too small. Figure 7 shows a flow velocity profile in a small-diameter pipe or duct as compared to a profile in a large-diameter pipe or duct.

Turbulence

If the surface over which the air passes is rough enough, disturbances are set up in the airflow so that it is no longer laminar but *turbulent*. Turbulent flow is not desirable — energy is lost since work has to be done to cause the turbulent or eddying flow.

Supersonic Flow

The various topics in this book will be developed on the basis of subsonic flow (slower than the speed of sound). The subjects of transonic and supersonic flow—airplane performance at speeds approaching and exceeding the speed of sound—are vast and beyond our scope here.

The behavior of air changes as the speed of sound is approached and may best be illustrated by considering *pressure waves.*

At low speeds, the flow of air around an airfoil is affected by pressure exerted on the airfoil, but it forms a streamline pattern around it, as seen in Fig. 8.

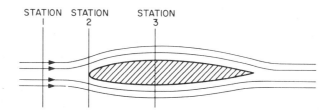

Fig. 8 Airflow around Symmetrical Airfoil–Subsonic Speed

The pressures at stations 2 and 3 in Fig. 8 affect the pressure in the airstream at station 1 somewhat ahead of the airfoil. Any change in pressure at the airfoil is almost instantly telegraphed ahead by *pressure waves* created by the change. The maximum speed at which a pressure wave travels is that of the speed of sound.

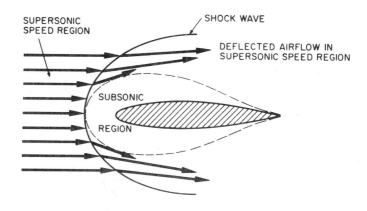

Fig. 9 Airflow around Symmetrical Airfoil–Supersonic Speeds

Supersonic Flow (cont'd)

At and above the speed of sound the situation is different. The pressure waves cannot be transmitted through the air fast enough, and the "telegraph" doesn't function. Accordingly, the airflow is not affected ahead of the airfoil, but hits it sharply. This creates an entirely different flow pattern, setting up a *shock wave* with different pressures ahead of and behind the wave. For an airfoil, the regions of various pressures are shown in Fig. 9.

Mach Number

Since the velocity of sound enters into considerations of airflow at high speeds, it is convenient to use the ratio V/c—where V is the airstream velocity and c is the speed of sound—as the criterion. This ratio is named the Mach number in honor of the Austrian physicist who first pointed out the relationship of pressure waves in a medium to the velocity of sound in the same medium. The velocity of sound in air depends on its absolute temperature. Accordingly, the velocity of sound in air is lower in extremely cold air than in warm.

Review Questions

1. State the equation of continuity.
2. A pipe 120 feet long tapers from a diameter of ten feet at one end to five feet at the other. What is the velocity at the exit of a gas entering:
   (a) the small end at 20 feet per second?
   (b) the large end at 20 feet per second?
3. Define a streamline and a stream tube.
4. Identify the three kinds of energy on which Bernouilli's equation is based.
5. Show how Bernouilli's Theorem is used to explain the difference in pressures on an airfoil.
6. What is laminar flow?
7. What is a boundary layer?
8. What is the effect of turbulence?
9. How do subsonic and supersonic flow around an airfoil differ?

MOTION

Laws of Motion

A number of basic laws deal with motion and the forces that generate or govern it. Three fundamental ones, enunciated by Sir Isaac Newton, are:

1. Every body remains at rest or moves in a straight line at a constant speed unless acted upon by an outside force.

2. When an outside force acts on a body, the changes are proportional to the magnitude of the impressed force and the mass of the body, and these changes take place in the line of direction of the impressed force, and depend upon the time during which the force acts.

3. To every action there is an equal and opposite reaction.

Since aerodynamics deals with the forces generated by air in motion, these *Laws of Motion* form the basis of all the equations in which forces are involved.

The first law, sometimes called the *Law of Inertia*, is easy to accept and understand. When a body is moving at a constant or uniform speed, in a straight line, it will increase or decrease in speed or be deflected from its original line of motion only if force is impressed on it. For an airplane, the impressed force could be a sudden change in the thrust supplied by the power plant, or the forces generated when air is deflected by moving the controls.

The second law is useful in considering the effect of forces. If the body was originally in motion, the new direction will be the resultant of two components: the original line of motion is one component, and the other is the line of action of the impressed force. This law leads to a useful equation regarding the magnitude of the force, namely,

$$F = Ma$$

where F is the impressed force
 M is the mass (W/g)
 a is the acceleration of the body caused by the impressed force

Laws of Motion (cont'd)

The third law is useful in both static and dynamic conditions. For example, a box weighing 100 pounds resting on a floor would have a reaction exactly equal to 100 pounds, or the carrying load of the floor. (The floor pushes up on the box with a force of 100 pounds.) This third law is useful in considering forces acting on an airplane. For example, if an airplane is flying at a constant altitude at a constant speed, the thrust force of the power plant will be exactly equal to the total resistance, or drag, of the airplane.

Wind, Motion, Velocity, and Speed

The science of aerodynanics deals with the behavior of moving air and the forces generated by the interaction of a moving body in still air, or a flowing stream of air on a stationary body, or a moving body in a moving stream of air.

A factor common to these three cases is that of *relative motion*, which is defined as the velocity of the air with reference to the solid body, be it a wing, an airplane, or any other object. Instead of *relative motion*, the term *relative velocity* is used whenever speeds are specified. The terms *relative wind, relative motion, relative speed,* and *relative velocity* are used interchangeably, though the first two terms refer more to the condition of the airflow, and the latter two terms refer more to quantitative measurements.

The air velocity is the average velocity of the air with relation to the object, but it is possible that the local air velocity over the object, at some particular point, may be lower or greater than the average. For example, the air moving over a curved object may be greater at the point of greatest curvature while the air velocity meeting the object is not so great. Similarly, the air entering an air-intake of an engine may be of a lower velocity than that of the air moving past the aircraft itself. These local velocities are significant when studying local pressures and airflows.

Speed Regimes

The operating speeds of aircraft vary from zero to beyond the speed of sound. The operational behavior of the aircraft varies greatly within the variou *speed regimes.* It is useful to classify them in terms of the *Mach number (M).* This number is the ratio of the air speed to the speed of sound under the same atmospheric conditions. (A plane flying at Mach 1

Speed Regimes (cont'd)

would be flying at exactly the speed of sound.) Accordingly, it is common practice to define these flight regimes more or less as follows:

Subsonic	M less than 0.75
Transonic	M greater than 0.75 but less than 1.20
Supersonic	M greater than 1.20 but less than 5.00
Hypersonic	M greater than 5.00

At the present time, flight speeds of practical aircraft are not higher than about Mach 2, so that the higher Mach numbers are of interest primarily in other fields.

Review Questions

1. An airplane is observed flying at a speed of 150 miles per hour on a calm day by an observer on the ground. What is the relative velocity of the airplane with respect to the ground?
2. An airplane is observed flying at a speed of 150 miles per hour by an observer on the ground, at a time when the wind velocity is 30 miles per hour in a direction opposite to that of the airplane. What is the relative speed of the airplane with respect to the air? What is the relative speed of the air with respect to the ground?
3. An airplane is flying at a speed of 200 miles per hour according to the air speed indicator in the cockpit. It is flying in a tail jet stream of 100 miles per hour. What is the speed of the airplane with relation to the air around it? What is its speed with relation to the ground?
4. State the three laws of motion.

AIRFOIL GEOMETRY

The *airfoil* is the cross section or the profile of the lift producing medium (wing) of the aircraft. It evolved over about sixty years from the early flat "plates" used for gliders to the present highly sophisticated curvatures of both the top and bottom surfaces of the wing. Since the design of the airfoil and the wing (these two terms are often used interchangeably by the layman) is important, it is necessary to learn the terminology.

Properly, the airfoil is the envelope of the *cross section of the wing*. Early flight experimenters used a wing which was essentially a flat panel with a fabric covering over the top of a structural framework. They soon found, however, that covering the bottom surface increased the lift effectiveness of the wing, and that a curvature of the panel was still better. Thus the cross section (airfoil) may differ at various parts of the wing.

Chord Line

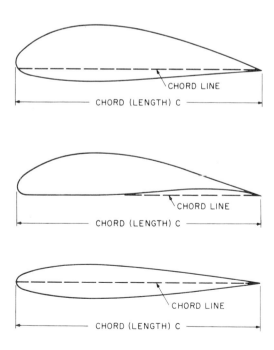

Fig. 10 Chord Designations—Typical Airfoils

Chord Line (cont'd)

The *chord line* or *chord* defines the length of the airfoil. It is a line drawn from the leading edge (front) to the trailing edge (rear). It may fall within or without the airfoil (Fig. 10). It may be so chosen as to make it a ready reference axis for indicating various ordinates of the airfoil. It is a useful dimension and an easily recognizable reference.

The terms chord line and chord are often used interchangeably, with chord in most common usage. The chord is usually designated by the lower-case letter "c".

Camber

The upper and lower surfaces of an airfoil are known as the *upper* and *lower cambers* (Latin: *camur*, curve), respectively. The distance halfway between the upper and lower cambers is called the *mean camber line*. Strictly speaking, an airfoil, being a cross section, has no upper or lower surface, so that it would be more nearly correct to say the upper camber line and the lower camber line, though it is common to just say the upper or lower camber.

When the wing itself is described, instead of the upper camber, upper surface is used; similarly, lower surface is used instead of lower camber.

The mean camber line (Fig. 11) is of interest to the aerodynamicist, since by varying the curvature of that line, aerodynamic properties of the wing are affected. It is also a basis for classifying certain airfoil shapes.

The *maximum camber* is the maximum displacement or thickness of the mean camber line from the chord line. The maximum camber and its location are expressed in a fraction or a percentage of the chord

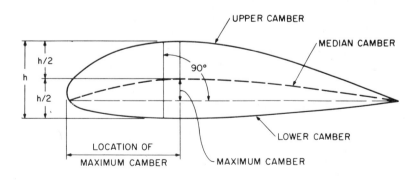

Fig. 11 Camber Designations for Airfoil

Camber (cont'd)

length. For airfoils useful at subsonic speeds, the maximum camber may vary from zero, for symmetrical airfoils, to 4 percent, and it is usually located at about 30 percent of the chord.

Leading Edge Radius

This term is used when constructing the airfoil and locating the *leading edge*. The center of the circle described by the leading edge radius is located on the mean camber line. Its coordinates may be referred to another reference line (Fig. 12).

The leading edge of the airfoil need not necessarily be an arc of a circle; some airfoils have sharp or pointed leading edges.

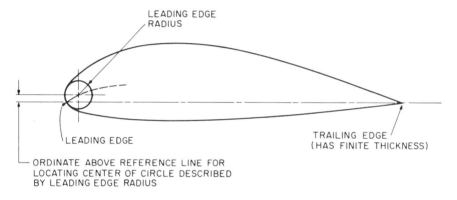

Fig. 12 Constructing the Leading Edge

Trailing Edge

The *trailing edge* is the rearmost edge of an airfoil where the top or upper camber line and the lower camber line intersect. In the actual wing, the trailing edge is not a knife-edge but it has some finite thickness, consistent with the materials used. The term can be used to refer to the rear portion of the airfoil or wing.

Thickness

The distance between the upper and lower cambers is the *thickness*, and it varies along the chord. The maximum thickness and its location are

Thickness (cont'd)

important reference values. There has been considerable experimentation with the maximum thickness and its location. Practical values for maximum thickness ratios lie between 6 and 18 percent of the chord length. The point of maximum thickness of an airfoil may be located at any distance between 30 and 40 percent of the chord. Different aerodynamic characteristics, not to mention structural considerations, are affected by changing the maximum thickness ratio and its location.

A wing might well have an airfoil of 15 to 18 percent maximum thickness ratio near the root with an airfoil of 6 percent thickness at the tip.

The point of maximum thickness ratio (Fig. 13) is usually located at 30 percent of the chord of the wing.

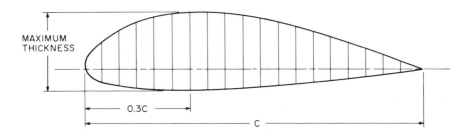

Fig. 13 Location of Maximum Thickness

Airfoil Coordinates

The airfoil coordinates are to be found in various NASA [National Aeronautics and Space Administration, formerly the National Advisory Committee for Aeronautics (NACA)] reports available from the U.S. Government Printing Office, Washington, D.C. A typical listing of the necessary data to construct an airfoil profile is shown in the accompanying table, Airfoil Coordinates.

These coordinates are given in percentages of the chord length. Therefore, to construct the contours of an airfoil, one would proceed as follows:

1. Lay out a horizontal line representing the x-axis and coinciding with the chord of the wing. On this line, mark the length of the chord in inches. To illustrate the use of the coordinates, assume a chord length of 50 inches.

Airfoil Coordinates (cont'd)

AIRFOIL COORDINATES

Station (percent)	Upper Ordinate (percent)	Lower Ordinate (percent)
0.00	0.000	−0.000
0.50	0.970	−0.870
0.75	1.176	−1.036
1.25	1.491	−1.277
2.50	2.058	−1.686
5.00	2.919	−2.287
7.50	3.593	−2.745
10.00	4.162	−3.128
15.00	5.073	−3.727
20.00	5.770	−4.178
25.00	6.300	−4.510
30.00	6.687	−4.743
35.00	6.942	−4.882
40.00	7.068	−4.926
45.00	7.044	−4.854
50.00	6.860	−4.654
55.00	6.507	−4.317
60.00	6.014	−3.872
65.00	5.411	−3.351
70.00	4.715	−2.771
75.00	3.954	−2.164
80.00	3.140	−1.548
85.00	2.302	−0.956
90.00	1.463	−0.429
95.00	0.672	−0.040
100.00	0.000	−0.000

Leading edge radius 1.00 percent   Slope of radius 0.084

2. Mark the stations as follows:

Station	Distance
0.00%	0.000 in.
0.50%	0.250 in.
0.75%	0.375 in.
1.25%	0.625 in.
2.50%	1.25 in.
etc.	etc.

3. Drop perpendiculars to the chord line at each of the stations, extending them above and below the chord line.

Airfoil Coordinates (cont'd)

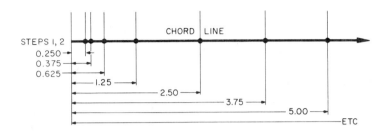

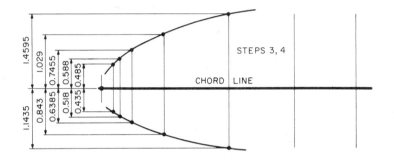

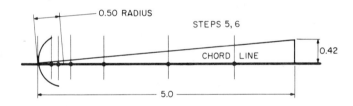

Fig. 14 Establishing the Airfoil Contour

4. On the perpendicular lines, mark the upper and lower ordinates in inches as indicated below:

Station	Upper Ordinate		Lower Ordinate	
0.000 in.	0.000%	0.000 in.	0.000%	0.000 in.
0.250 in.	0.970%	0.485 in.	0.870%	0.435 in.
0.375 in.	1.176%	0.588 in.	0.036%	0.518 in.
0.625 in.	1.491%	0.746 in.	1.277%	0.639 in.
etc.	etc.	etc.	etc.	etc.

Airfoil Coordinates (cont'd)

5. For the airfoil in question, the slope of the radius is given as 0.084. At the zero station, erect a tangent with a slope of 0.084. This value is a ratio, not a percentage. To obtain the tangent, choose any convenient base length, say 5 inches, erect a perpendicular at the end of that base line, and measure a distance upward equal to five times 0.084 or 0.42 inch. Draw a line from the zero station to the 0.42-inch point on the perpendicular. This is the radius line.
6. The radius of the leading edge circle is given as 1 percent of the 50-inch chord length, or 0.50 inch. Draw a circle of this radius with the center on the radius line 0.50 inch from the leading edge.

These steps are illustrated in Fig. 14. Although shown in separate drawings, in actual practice all the steps would be carried out with one base line.

Review Questions

1. Define chord, camber, leading edge radius, leading edge, and trailing edge.
2. How is the thickness ratio of an airfoil expressed?
3. A rectangular wing has an airfoil with a thickness ratio of 6 percent at the tip, and 18 percent at the root. What is the thickness ratio of the airfoil halfway between the tip and the root?
4. What are the usual limits for thickness ratios of airfoils?
5. What is meant by "airfoil geometry"?

WING GEOMETRY

All the airfoil geometry applies to the wings as well, since the airfoil is the cross section or profile of the wing. There may be an infinite number of airfoil sections for a wing when different airfoils are used for the root and tip stations. However, other geometrical features of the wing are important in the aerodynamics of flight.

Span

The horizontal distance between the wing tips is the *wing span*, or span. It is usually expressed in feet for the completed airplane, but for construction purposes and manufacturing layouts, it is given in inches.

Root

The *root* of the wing is located halfway between the wing tips, or at mid-span. It is a useful reference plane for establishing certain construction data. Sometimes the root of the wing may be considered to be located at the side of the fuselage.

Chord

The *chord,* measured in feet or inches, of any airfoil of the wing is also the chord of the wing where the airfoil is located. For a *rectangular* planform of a wing, the chord is the same from wing tip to wing tip. For a tapered planform, the chord obviously varies.

As shown in Fig. 15, for a wing tip which is not rectilinear, the tip chord may be located at the extreme tip or slightly inboard. The arbitrary wing tip length may have been calculated from considerations of the mean geometric chord (see later). Or, the airfoil sections between the geometric chord and the root chord have been established by the relationship of the airfoil ordinates of the tip chord section to that of the root chord section.

Mean Geometric Chord

In studying longitudinal stability in flight, it is useful to consider a specific chord length known as the *mean aerodynamic chord,* at which the results of all the aerodynamic forces may be assumed to be concen-

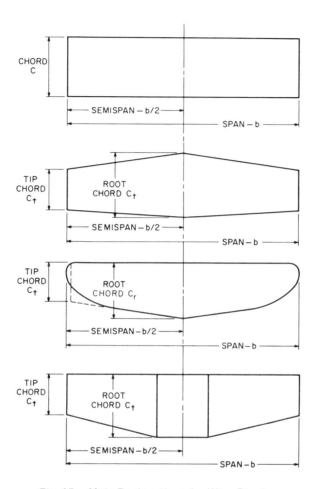

Fig. 15 Main Designations for Wing Planform

trated. It is the chord location of only one half of the wing, either up to the side of the fuselage, for a wing whose center portion is blanketed by the fuselage, or up to the plane of symmetry if not so blanketed.

For preliminary calculations, and for most practical purposes, the *mean geometric chord* (MGC) may be used instead. The assumption here is that each unit of area has on it the same aerodynamic forces as every other unit. Under this assumption, the geometric chord would be at the balance point.

For a rectangular wing, the mean geometric chord would be located just halfway along that portion of the semi-span which extends from the fuselage.

Mean Geometric Chord (cont'd)

For a tapered or trapezoidal wing, the mean geometric chord length and location can be determined graphically or calculated. To determine these figures graphically, refer to Fig. 16.

1. Lay out the planform of the wing to a convenient scale.
2. Extend the root chord (in either direction) for a distance equal to the length of the tip chord.
3. Extend the tip chord, in the direction opposite to that used for the root chord, for a distance equal to the root chord.
4. Connect the tips of the extensions made in steps 2 and 3.
5. Connect the midpoints of the root and tip chords.
6. The mean geometric chord line will be located at the intersection of the two lines drawn in steps 4 and 5.
7. Draw a line parallel to the root chord at the intersection of the lines indicated in step 6. The length of the mean geometric chord is the distance between the leading and trailing edge of the wing at this point.

The mean geometric chord can also be determined mathematically. By taking moments of the areas about the root chord and dividing those moments by the area, the location of the mean geometric chord (MGC) outboard can be found. The trapezoid can be divided into a rectangle and a triangle. Referring to Fig. 17a, the center of gravity of the rectan-

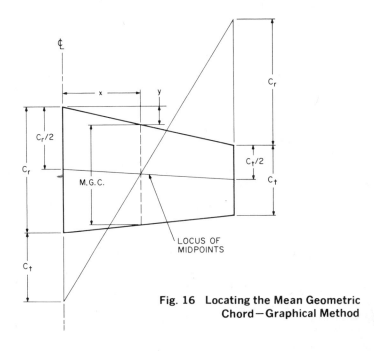

Fig. 16 Locating the Mean Geometric Chord—Graphical Method

Mean Geometric Chord (cont'd)

gular portion would be at a distance of $b'/2$ outboard. The MGC of the rectangular portion would be located at this point. The area of the rectangle is $b'C_t$; the moment of this area about the root chord as a base would be

$$(b'C_t)\frac{b'}{2}$$

The center of gravity of the triangular part would be at a distance of $b'/3$ outboard, and the moment of the triangular area would be

$$\left[\frac{b'}{2}(C_r - C_t)\right]\left[\frac{b'}{3}\right]$$

where $(b'/2)(C_r - C_t)$ is the area of the triangle

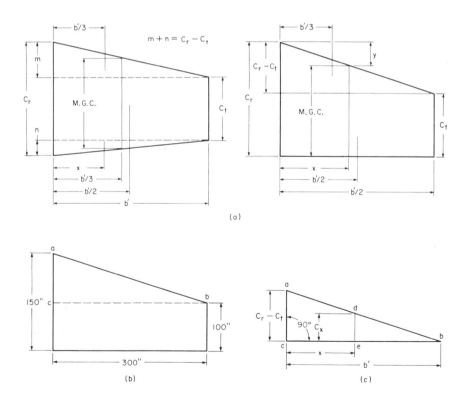

Fig. 17 a. Determining the MGC—Reference Planforms for Mathematical Method   b. Reference Diagram for Numerical Example   c. Reference Diagram for Determining C_x

Mean Geometric Chord (cont'd)

Adding these two moments, and dividing the combined area gives the location outboard of the MGC, or

$$x = \frac{(b'C_t)\frac{b'}{2} + \left[\frac{b'}{3}\right]\left[\frac{b'}{2}(C_r - C_t)\right]}{b'C_t + \frac{b'}{2}(C_r - C_t)}$$

This reduces to

$$x = \frac{b'}{3}(C_r + 2C_t)/(C_r + C_t)$$

The length of the MGC can be found by making use of the relationship existing between corresponding sides of similar triangles:

$$MGC = \frac{2}{3}(C_r + C_t^2)/(C_r + C_t)$$

Example: Determine the length of the mean geometric chord and its location for the wing planform shown in Fig. 17b. For the diagram, the available data are $b' = 300$ in., $C_t = 100$ in., $C_r = 150$ in. It is necessary to find the MGC.

Solution: The MGC is located some distance from the root chord. In the diagram it is measured to the right.

Remember that the MGC is considered to be the equivalent of the MAC, at which chord all the aerodynamic forces are assumed to be located. The aerodynamic forces would have, therefore, the same effect if they were distributed uniformly over the entire wing.

To determine the distance, x, one takes moments about the root chord, assuming a uniform load, let's say, of w pounds per square inch. (Its value need not be known.)

For the rectangular portion: the load over the rectangular portion of the wing, whose area is $C_t b'$, would be $C_t b'w$. Numerically, this would be equal to (100)(300)w or 30,000 w pounds. The center of gravity of this load would be halfway to the right of the base or root chord or $b'/2$. (Recall that the center of gravity of a rectangle is at its center.)

The moment of the load over the rectangular portion about the base or root chord would be

$$M_1 = (C_t b'w)\left(\frac{b'}{2}\right) = C_t(b')^2 \frac{w}{2}$$

$$= (100)(300)^2 \frac{w}{2} = 9,000,000 \frac{w}{2}$$

$$= 4,500,000 \text{ w inch-pounds}$$

WING GEOMETRY 33

Mean Geometric Chord (cont'd)

The load over the triangular portion of the wing, whose area is $(C_r - C_t)b'/2$, would be $(C_r - C_t)b'w/2$. Numerically, this would be equal to

$$(150 - 100)\left(\frac{300}{2}\right) w = 7500 \text{ w pounds}$$

The center of gravity of the triangle is located at a distance one-third of the altitude up from its base or $b'w/3$. The moment of the load over the triangular portion about the base would be

$$M_2 = (C_r - C_t)\left(\frac{b'}{2}\right)(w)\left(\frac{b'}{3}\right) = (C_r - C_t)\left(\frac{w}{6}\right)(b')^2$$
$$= (150 - 100)\left(\frac{w}{6}\right)(300)^2$$
$$= 750{,}000 \text{ w inch-pounds}$$

For the whole wing, the total load over the wing is the sum of the rectangular and triangular portions, or

$$L = C_t b' w + (C_r - C_t)\frac{b'}{2} w$$

or numerically,

$$L = 30{,}000w + 7500 = 37{,}500 \text{ w pounds}$$

or algebraically,

$$L = \frac{b'w}{2}(2C_t + C_r - C_t) = \frac{b'w}{2}(C_t + C_r)$$

The total moment over the wing would be

$$M = M_1 + M_2 = \left[C_t(b')^2\frac{w}{2}\right] + \left[(C_r - C_t)\left(\frac{w}{6}\right)(b')^2\right]$$

or numerically,

$$M = 4{,}500{,}000w + 750{,}000w$$
$$= 5{,}250{,}000w \text{ inch-pounds}$$

or algebraically,

$$M = \frac{(b')^2}{6}(w)(2C_t + C_r)$$

Mean Geometric Chord (cont'd)

The distance x is found by dividing the moment of all the load by the total load or

$$x = \frac{M}{L} = \left[(b')^2 \frac{w}{6}(2C_t + C_r)\right] \div \left[\frac{b'w}{2}(C_t + C_r)\right]$$

$$= \frac{b'}{3}(2C_t + C_r) \div (C_t + C_r)$$

or numerically,

$$x = \frac{M}{L} = \frac{5,250,000w}{37,000w} = 140 \text{ inches}$$

The value of x is measured up from the base about which the moments were measured.

What is the length of the MGC? Its location has been found, and it is now necessary to resort to some geometric principles since a triangle is involved. Consider the triangle of Fig. 17c. The triangles abc and dbe are similar, since all interior angles are equal; thus,

$$\frac{C_x}{C_r - C_t} = \frac{b' - x}{b'}$$

which is read: "the base C_x is to base $(C_r - C_t)$ as the altitude $(b' - x)$ is to altitude b'," or, solving for C_x,

$$C_x = \left(\frac{b' - x}{b'}\right)(C_r - C_t)$$

Substituting $b' = 300$, $x = 140$ in., $C_r = 150$ in., $C_t = 100$ in., the numerical value for C_x becomes

$$\left(\frac{300 - 140}{300}\right)(150 - 100) = 26.66 \text{ inches}$$

Substituting for x, to obtain an algebraic expression,

$$C_x = \frac{1}{b'}\left[b' - \frac{b'}{3}\frac{(2C_t + C_r)}{C_t + C_r}\right](C_r - C_t)$$

$$= \frac{(C_t + 2C_r)(C_r - C_t)}{3(C_t + C_r)}$$

Mean Geometric Chord (cont'd)

Again, substituting for C_t and C_r, it will be found that $C_x = 26.66$ inches. The chord of the rectangular portion is C_t throughout; hence,

$$MGC = C_t + C_x$$

$$= C_t + \left[\frac{(C_t + 2C_r)(C_r - C_t)}{3(C_t + C_r)}\right]$$

$$= \frac{2}{3}\left(C_r^2 + C_r C_t + C_t^2\right) / \left(C_t + C_r\right)$$

Substituting the values $C_t = 150$ in., and $C_t = 100$ in.,

$$MGC = \frac{2}{3}\left[\frac{(150)^2 + (150)(100) + (100)^2}{100 + 150}\right]$$

$$= 126.66 \text{ inches}$$

Problems

1. Verify the formula for determining the length of the mean geometric chord.
2. A trapezoidal wing has a root chord of 100 inches and a tip chord of 50 inches. Its semi-span is 200 inches. Locate the mean geometric chord and determine its length by graphical means. Compare the results obtained by using the formulas.

Planforms

As already shown in Fig. 15, the wing may have an almost infinite variety of planforms. Generally, the wings are trapezoidal with rounded wing tips—not only for aerodynamic reasons but for structural reasons as well.

The airfoil for the root chord has a thickness ratio of 12 to 18 percent, while the tip chord has a thickness ratio of six to nine percent. The thinner airfoil is desirable for better overall aerodynamic effectiveness, but a thicker airfoil is required for structural reasons. Since the bending moment of the wing is greatest at the root, it is necessary to have a greater wing depth there for efficient structural design.

Angles

The wing planform may be swept forward or backward. When viewed head-on, it may be noted that the wing tip may be higher than the wing portion at the root. Viewed from the side, the wing as a whole may seem to slope with reference to the main longitudinal axis of the airplane. These features are built-in angles.

Sweepback

The *sweepback* angle of the planform of the wing is designated by the capital Greek letter lambda, Λ. It may be measured (Fig. 18) between the line perpendicular to the plane of symmetry of the airplane and the leading edge of the wing; or along a line representing the locus of the quarter chord points of the wing; or along a line representing the locus of the aerodynamic centers of the wing (to be described later when considering aerodynamic characteristics).

The sweepback of a wing affects its maximum lift coefficient (generally reducing it) and its stall characteristics. At transonic and supersonic speeds, it ameliorates the compressibility effects.

For supersonic designs, a sharply swept-back wing, presenting an arrowhead outline (Fig. 19), and called a "delta" wing (since the planform resembles the Greek letter delta), has several desirable aerodynamic characteristics—notably lower drag coefficients.

Sweepback is also useful for subsonic aircraft. It meets certain longitudinal—and to some degree directional—stability requirements in flight.

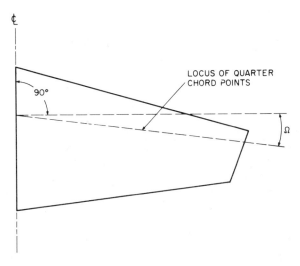

Fig. 18 Measuring the Sweepback Angle

WING GEOMETRY

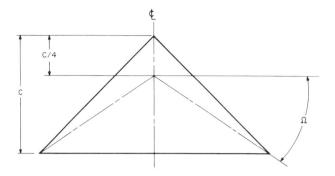

Fig. 19 A Delta Wing Planform

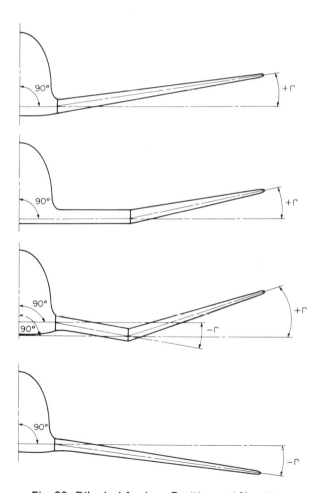

Fig. 20 Dihedral Angles—Positive and Negative

Dihedral

The upward slope of the wing when viewed from the wing is known as *dihedral*. The angle is positive when the wing slopes upward, and negative when it slopes downward. The *dihedral angle* is usually designated by Γ (capital Greek letter gamma). Figure 20 shows some positive and negative dihedral angles.

The angles have been exaggerated in the diagrams. Positive dihedral angles vary up to six degrees. Negative dihedral is used to offset the effects of large sweepbacks.

The angle is measured between the plane drawn perpendicular to both the plane of symmetry and the plane determined by the wing chords. Since the wing may be twisted about its spanwise axis, the plane is passed through the quarter point of the mean geometric chord. A wing twisted about its spanwise axis is said to have an *aerodynamic twist* since the angle of incidence of each successive cross section is different.

Angle of Incidence

The positioning of the wing with relation to the fuselage may be conditioned by structural requirements or by aerodynamic considerations. The angle may be referred to any convenient reference axis determined by the designer. The *angle of incidence* (Fig. 21) is the angle located between the chord line of the mean geometric chord and the longitudinal axis.

The angle of incidence of the root and tip sections of a wing need not be the same. A wing with different angles at those stations would have changes in angles all along the span.

Similarly, the horizontal tail surfaces may have an angle of incidence with relation to the accepted reference line.

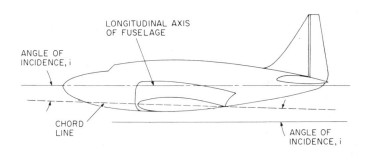

Fig. 21 The Angle of Incidence

Wing Area

The area of the wing is the area included in the planform projected upon the plane of the chords. Even if the wing chords from root to tip have different angles with relation to each other, so that the plane through the chords would be warped, the plane of the wing may still be assumed to be flat without any serious error. The wing area is generally designated by the letter S, and is expressed in square feet. (The area covered by engine nacelles and the fuselage is included in the wing area, as indicated by the shaded areas in Fig. 22.)

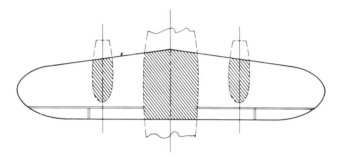

Fig. 22 Effective Wing Area, as Shown by Solid Outline

Aspect Ratio

Early in aerodynamic history, it was discovered that the aerodynamic characteristics of an aircraft depended upon the ratio of the wing span to the wing chord. This ratio is known as the *aspect ratio* (AR), and is defined by the equation

$$AR = \frac{(\text{span})^2}{\text{area}} = \frac{b^2}{S}$$

For a rectangular wing, this equation reduces to

$$AR = \frac{b^2}{bC} = \frac{b}{C}$$

For a glider or sailplane, the aspect ratio may be as high as 35. Commercial airplanes have wings of aspect ratios between five and eight, while jet fighters that operate at transonic and low supersonic speeds have wings of aspect ratios near 3.5. Figure 23 shows three possible planforms for an airplane wing. All the wings have equal spans, but the aspect ratio of each may be different if the areas are different.

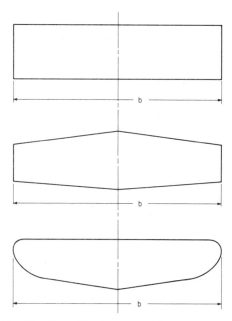

Fig. 23 Wing Planforms with Equal Spans

Review Questions

1. What is the span of a wing? A chord? The aspect ratio?
2. A wing has an area of 200 square feet and a span of 30 feet. What is its aspect ratio?
3. A trapezoidal wing has a root chord of 12 feet, and a tip chord of 6 feet. What is the angle of sweepback when:
   (a) The leading edge of the wing is perpendicular to the plane of symmetry?
   (b) The trailing edge of the wing is perpendicular to the plane of symmetry?
   (c) The locus of the quarter points of the chords is perpendicular to the plane of symmetry?
4. Determine graphically the mean geometric chord in question 3(c).
5. What are these angles: (a) incidence, (b) dihedral, (c) sweepback?
6. Calculate the aspect ratio of each of the following wings:
   (a) A rectangular wing with a span of 40 feet and a chord of 8 feet.
   (b) A trapezoidal wing (each half is a trapezoid) with a span of 40 feet, a tip chord of 6 feet, and a root chord of 10 feet.
   (c) A trapezoidal wing with a span of 40 feet, a tip chord of 4 feet, and a root chord of 12 feet.

REFERENCE AXES

Since the airplane has the dimensions of length, width, and height, and the ability to move upward, forward, and sidewise, it has three main axes. These form a convenient reference system for calculations.

Body Axes

A system of three mutually perpendicular axes passing through the center of gravity with the X-axis parallel to some longitudinal reference (such as the thrust axis, the mean geometric chord, the mean aerodynamic chord, or some axis of the fuselage such as an axis of symmetry) constitutes a *body axes* system. Such a system is shown in Fig. 24. The X-Z plane is in the plane of symmetry of the airplane.

The positive axes are shown as solid lines with the Z-axis positive downward, the Y-axis positive toward the right when looking forward to the nose of the airplane, and the X-axis positive forward.

All forces and straight-line motions acting in the direction of these positive axes are positive; those in the opposite direction are negative. The notation of axes as given need not always be adhered to if it is more convenient to do otherwise. For example, in problems dealing with forces in the X-Z plane, it sometimes seems easier to use an X-Y coordinate system. Similarly, it will be noted that the positive X-Z quadrant corresponds to what might be considered the third quadrant instead of the first, when viewing the aircraft's plane of symmetry with the tail surfaces to the right.

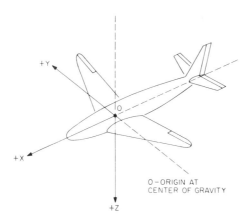

Fig. 24 Body Axes of an Airplane

Body Axes (cont'd)

In addition to forces and straight line motions, angular motion or rotation, as well as moments, will be considered. These are considered if, in looking along the positive axes (with arrows pointing in the positive direction), they act in a clockwise direction.

Wind Axes

Another set of three mutually perpendicular axes, with the center of gravity as the origin, is known as the wind axes with the X-axis parallel to the direction of the relative wind, as shown in Fig. 25. The X-axis is considered positive aft instead of positive forward. The Y-axis remains positive as for the body axes, while the Z-axis is positive upward. In spite of the different conventions used, there should be no confusion as long as the system employed is duly noted.

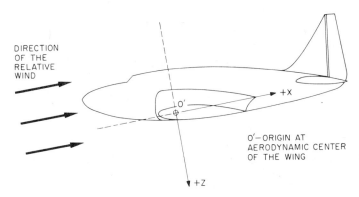

Fig. 25 Wind Axes

Review Questions

1. What are the body axes of an airplane?
2. What are the wind axes of an airplane?
3. When are these axes used?
4. What are designations of the three axes?
5. Which is the positive quadrant in each of the two systems?

BASIC AIRFOIL AERODYNAMICS

The *aerodynamic characteristics* of airfoils are a function of:

1. Airfoil shape, thickness, and surface texture.
2. Wing geometry, such as taper and aspect ratios, aerodynamic twist, sweepback, dihedral, etc.
3. Compressibility effects.
4. Scale effects.
5. Angle of attack.

Angle of Attack

The *angle of attack* is the angle measured between the line of relative wind and a given reference line. For an airfoil, the accepted reference line is the chord line. (See Fig. 26.) For a fuselage, it would be any fixed line marked by the manufacturer or designer as the reference line. For symmetrical objects, the axis of symmetry is often a useful reference line.

The angle of attack is usually designated by the Greek lower case letter α (alpha). Useful angles of attack in stable flight regimes are approximately between 0 and about 18 degrees for an airfoil.

Fig. 26 Measuring the Angle of Attack

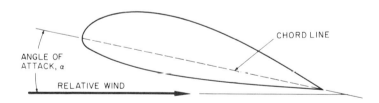

Aerodynamic Forces

The *aerodynamic forces* acting on the wing are caused by the *relative wind* passing over its surface. The forces are negative pressure (suction) and positive pressure forces. These local pressures act perpendicularly to the surface and may be measured experimentally by connecting tiny orifices in the surfaces to liquid manometers, as shown in Fig. 27.

The pressure distribution over the airfoil changes with angle of attack. The intensity of the pressure at any point is a direct function of the rela-

Aerodynamic Forces (cont'd)

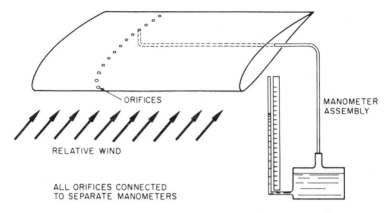

Fig. 27 Pressure over an Airfoil—Experimental Measuring Setup

tive wing speed. The pressure distribution is unique for each object, in that it is dependent upon the shape of the object over which the pressure is measured.

The pressure at any point in Fig. 28 is indicated by a vector (an arrow pointing in the direction of the pressure, or force, and drawn to a suitable scale—with 1 inch representing 100 pounds, for example). These vectors represent pressures in pounds per square inch. When a sufficient number of pressures along the surface have been measured and represented on the profile of the wing (in other words, the airfoil), a line is drawn connecting the ends of the arrows to form an *envelope*. If, for any reason, the pressure were desired at any point between those measured at given stations, they could be scaled from the drawing by determining the length of the perpendicular from the desired point to the intersection with the envelope.

The pressure envelopes for a symmetrical airfoil are shown for two different angles of attack.

Arrows pointing outward from the airfoil surface indicate suction, or negative pressure. The effect of this air pressure would be to pull the surface covering away from its supporting structure.

Arrows directed toward the surface indicate that the air pressure, a positive pressure, is pushing against the surface.

Since the airfoil is symmetrical, and the relative wind strikes the airfoil at a point directly parallel to the axis of symmetry, it will be noted that the pressure distribution is symmetrical. Accordingly, it may be surmised (if it is not apparent by inspection of Fig. 28) that the upward forces are directly counterbalanced by the downward forces. The net result would be no force nor lift perpendicular to the direction of motion.

Aerodynamic Forces (cont'd)

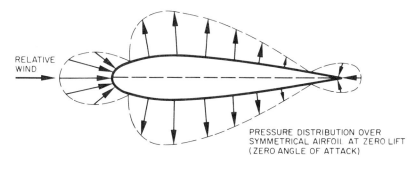

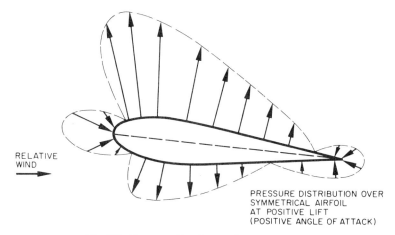

Fig. 28 Pressure Envelope—Symmetrical Airfoil

In the direction of the line of action of the relative wind, the net force is the resistance, or drag, that the object offers to the wind.

The pressure distribution over nonsymmetrical airfoils is similar, as shown in Fig. 29.

Resultant Force

All of these forces could be summed up to give one force or *resultant*, R. This force would be found to be proportional to the density of the air, w, the wing area, S, and the square of the speed, v. Symbolically, it could be written

$$R \approx w, S, v^2$$

where $\approx$ is the symbol for proportionality

46 FUNDAMENTAL OF AIRCRAFT FLIGHT

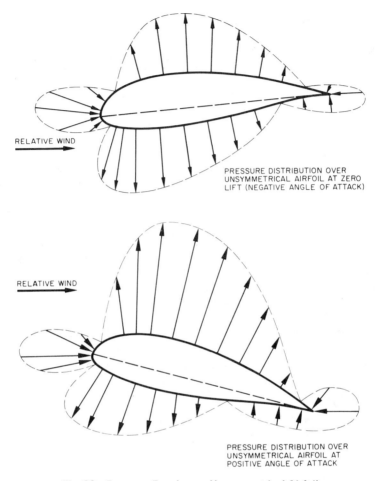

Fig. 29 Pressure Envelope—Unsymmetrical Airfoil

Resultant Force (cont'd)

The equation for kinetic energy, K.E., where m is mass, has the form

$$\text{K.E.} = \frac{1}{2}mv^2$$

Accordingly, the equation for the resultant force may be written

$$R = \frac{1}{2}\rho S v^2 k$$

where k is a constant that makes allowance for changes in shape of the object, changes in angle of attack, surface roughness, etc.

ρ (corresponding to m in the kinetic energy equation) is equal to w/g

BASIC AIRFOIL AERODYNAMICS 47

Lift and Drag Forces

The resultant force is not a convenient parameter to work with. Instead, two components, lift and drag, are far more useful, and are the quantities sought in experimental work.

The lift force, conveniently called lift, is that component of the resultant force perpendicular to the line of action of the relative wind. It is designated by capital L.

The drag force, called drag, designated by capital D, is the component parallel to the line of action of the relative wind.

Referring to Fig. 30, the resultant force is designated by the letter R. The component perpendicular to the line of action of the relative wind is the lift, or

$$L = R \sin \theta$$

The component parallel to the line of action of the relative wind is the drag, or

$$D = R \cos \theta$$

Example 1: The resultant force acting on a wing is 1200 pounds. Its line of action makes an angle of 55 degrees with the line of flight. What are the lift and drag of the wing? (Refer to Fig. 30.)

Solution: For the problem stated,

$$R = 1200 \quad \text{and} \quad \theta = 55 \text{ degrees}$$

Referring to a table of trigonometric functions,

$$\sin 55° = 0.8192$$
$$\cos 55° = 0.5736$$

Hence,
$$L = R \sin 55° = 1200 (0.8192) = 983 \text{ pounds}$$
And
$$D = R \cos 55° = 1200 (0.5736) = 688.3 \text{ pounds}$$

Example 2: The lift of a wing is 1200 pounds and the drag is 400 pounds. What is the resultant force? At what angle to the line of flight is it acting?

Solution: Since

$$L = R \sin \theta \quad \text{and} \quad D = R \cos \theta$$

Lift and Drag Forces (cont'd)

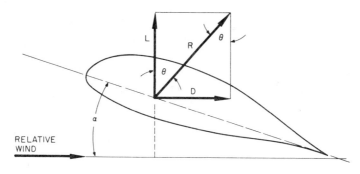

Fig. 30 Resolution of Forces

by squaring both sides of the two equations and adding, the following is obtained:

$$L^2 + D^2 = R^2\sin^2\theta + R^2\cos^2\theta$$
$$= R^2(\sin^2\theta + \cos^2\theta)$$

But,

$$\sin^2\theta + \cos^2\theta = 1$$

Hence,

$$R^2 = L^2 + D^2$$
$$R = \sqrt{L^2 + D^2}$$
$$= \sqrt{1200^2 + 400^2}$$
$$= 1264.9$$

Dividing the lift and drag equation results in

$$\frac{L}{D} = \frac{R\sin\theta}{R\cos\theta} = \tan\theta$$

Therefore,

$$\tan\theta = \frac{1200}{400} = 30$$

The table of trigonometric functions indicates that

$$\theta = 71.55 \text{ degrees}$$

Lift and Drag Equations

The equation for lift is expressed as follows:

$$L = \frac{1}{2}\rho v^2 C_L S \quad \text{or} \quad L = qC_L S$$

where L = lift, in pounds
ρ = mass density of the air
= 0.002738 slug per cubic foot at sea level for standard atmosphere
v = air velocity, feet per second
q = $(1/2)\rho v^2$, the dynamic pressure of the moving airstream, in pounds per square foot
S = wing area, in square feet
C_L = nondimensional lift coefficient

The equation for the drag is expressed as

$$D = \frac{1}{2}\rho v^2 C_D S \quad \text{or} \quad D = qC_D S$$

where D = drag (in pounds)
C_D = nondimensional drag coefficient

While C_L and C_D are nondimensional constants, they vary with angle of attack. It would not be worthwhile to develop an equation incorporating the effect due to the angle of attack. Charts, graphs, or tables are prepared to indicate the variation of these constants.

A quantity is designated as nondimensional when it does not involve the dimensions of mass (M), length (L), or time (T). The coefficients must be used in equations using consistent units for the other quantities.

Example 1: What is the dynamic pressure in an airstream having a velocity of 200 miles per hour under sea-level conditions?

Solution: The dynamic pressure, designated by the letter q, is

$$q = \frac{1}{2}\rho v^2$$

The v is expressed in feet per second, and at sea level, under standard air conditions, $\rho = 0.002738$ slugs per cubic foot. Hence, converting miles per hour to feet per second, multiply the number of miles by 5280 (the number of feet in a mile) and divide by 3600 (the number of seconds in an hour). The ratio of 5280 to 3600 is 1.467 feet per second, hence,

$$q = \frac{1}{2}(0.002378)(200 \times 1.467)^2 = 102 \text{ pounds per square foot}$$

Lift and Drag Equations (cont'd)

This example helps to illustrate the enormous pressure that air can exert at relatively high but subsonic speeds. Such a pressure, however, could only be exerted if the air were forced to come to a dead stop. Because air passes over an object, the pressures on the object are considerably reduced.

Example 2: An airplane with a wing area of 200 square feet is flying at 200 miles per hour, at a lift coefficient of 0.45 and a drag coefficient of 0.035 at standard sea-level conditions. What are the lift and drag?

Solution: The lift equation is

$$L = qC_L S$$

where $q = 102$ (as calculated in the previous equation)
$C_L = 0.45$
$S = 200$

Hence,

$$L = 102 (0.45) (200) = 9180 \text{ pounds}$$

The pressure per square foot is $9180/200 = 45.9$ pounds acting in the lift direction. The drag equation is

$$D = qC_D S = 102 (0.035) (200) = 714.0 \text{ pounds.}$$

The pressure per square foot is 3.57 pounds, acting in the drag direction.

Presenting Aerodynamic Data

Aerodynamic data, such as C_L or C_D, may be presented in tabular form or by graphical presentation. Both forms are useful.

The tabular form lists precise values which have to be available whenever dependent values have to be found by calculation. For this reason, the values of lift and drag coefficients are often given to at least three decimal places, for corresponding angles of attack.

Graphical presentation of data usually gives a better "picture" of what is happening: whether the drag coefficient is increasing or decreasing with angle of attack and how fast; whether the lift coefficient is dropping rapidly in value as the maximum lift coefficient is reached; etc. Also, when comparing the aerodynamic characteristics of airfoils, the graphical presentation of each should be placed on the same diagram for easier analysis.

Presenting Aerodynamic Data (cont'd)

In plotting aerodynamic characteristics of airfoils, it is customary to use the values of angle of attack (in degrees) as the abscissa, and to use the vertical scale, or ordinate, for other aerodynamic characteristics. Sometimes it may be useful to use the lift coefficient for the horizontal scale, or abscissa; or, for a different purpose, the drag coefficient may be used for the abscissa.

Graphical Representation

The aerodynamic characteristics of an airfoil (such as C_L and C_D) are usually plotted against the angle of attack in degrees, so that one curve can be obtained for the lift coefficient C_L and another for the drag coefficient C_D. These values are obtained from lift and drag forces, measured in the wind tunnel, for successive angles of attack.

In plotting the results, as in Fig. 31, the scale is designed so that the two resulting curves are within the confines of the paper used, and do not interfere with each other.

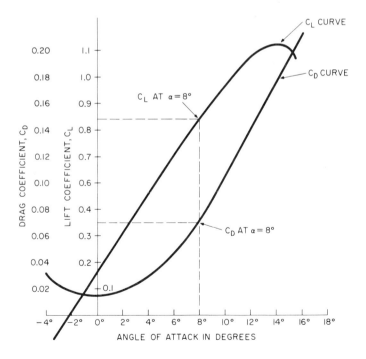

Fig. 31 C_L and C_D vs Angle of Attack—Graphical Representation

Graphical Representation (cont'd)

In addition to the two curves shown, a third curve could be drawn, showing the ratio of C_L to C_D (called L/D) for each angle of attack. This ratio has to be calculated since it is not a quantity measured directly by either model or full-scale tests.

Sometimes the coefficient C_L is plotted against the drag coefficient C_D with the former as the ordinate and the latter as the abscissa. Such a plot (Fig. 32) is sometimes called a "polar" diagram.

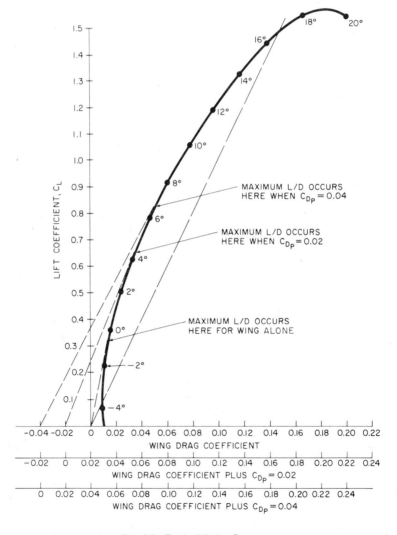

Fig. 32 Typical Polar Curve

BASIC AIRFOIL AERODYNAMICS

Graphical Representation (cont'd)

The plotting points are for the specific C_L and C_D values for a given angle of attack, and a tangent drawn from the origin to the C_L-C_D curve would determine the point of maximum L/D ratio; any other line cutting the curve would do so at two places, indicating that, other than at the maximum value of L/D, there are two possible values of C_L and C_D.

A polar diagram may have an advantage for studying the effects of parasite resistance on the value of L/D with relation to angle of attack. Adding C_{DP} to C_D has the effect of moving the origin of the abscissa to the left. Hence, if C_{DP} is plotted towards the left, and a tangent drawn from the new origin, it will be found that the L/D will occur at a greater angle of attack than for the wing alone, and that the greater the parasite drag coefficient, the greater the angle of attack at which the maximum L/D occurs.

Review Questions

1. How is the angle of attack measured?
2. What is a pressure envelope?
3. What is the resultant force?
4. State the lift and drag equations, identifying each term.
5. What is the drag of a wing whose lift is 900 pounds, while the resultant force on the wing is 1200 pounds? What angle does the resultant force make with the line of action of the relative wind?
6. How are lift and drag coefficients usually presented?
7. What can you say about the lift and drag coefficients with reference to: (a) dimensionality, and (b) variation with angle of attack?
8. What is the dynamic pressure of an air stream moving at 350 feet per second, under standard conditions?
9. What is the lift of a wing having a wing area of 300 square feet, flying at a lift coefficient of 0.65, under standard sea-level conditions, at a speed of 150 miles per hour?

ADVANCED AIRFOIL AERODYNAMICS

The Lift Coefficient

Lift coefficients that are useful in determining the performance of the airplane occur at angles of attack from about 0 degrees to about 18 degrees, for the range of aspect ratios used for aircraft wings operating in the subsonic range. The maximum lift coefficient usually occurs near 15 to 18 degrees

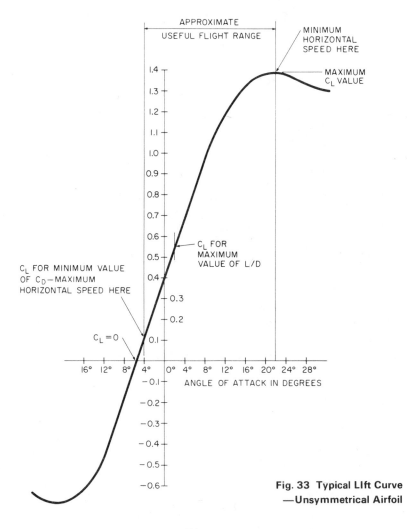

Fig. 33 Typical Lift Curve
—Unsymmetrical Airfoil

The Lift Coefficient (cont'd)

angle of attack. Beyond this angle, the maximum lift coefficient drops unless special lift increase devices are used. Figure 33 shows a typical variation of the lift coefficient with angle of attack.

For a symmetrical airfoil, the lift curve would pass through the origin for $C_L = 0$, and the curve in the third quadrant would be the same as that in the first quadrant. However, the curve shown is for an unsymmetrical airfoil, for which the maximum negative value of the lift coefficient would be less than the maximum positive value of the lift coefficient. (The various notations made on the curve will have more meaning after discussions of horizontal flight.)

By reducing the aspect ratio of the wing, the maximum lift coefficient is achieved at a higher angle of attack (Fig. 34).

The angle of maximum lift is also known as the "burble point." Here the flow tends to become turbulent.

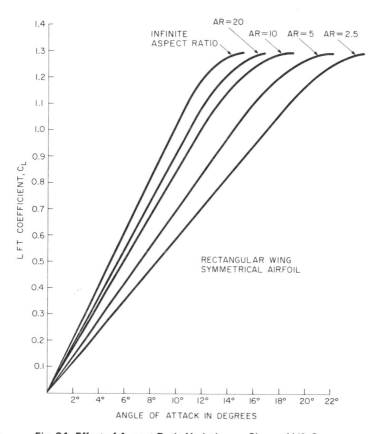

Fig. 34 Effect of Aspect Ratio Variation on Slope of Lift Curve

The Lift Coefficient (cont'd)

The maximum lift coefficient is a function of the thickness ratio for the same family of airfoils. Typical values are listed below.

Thickness Ratio	Maximum Lift Coefficient
6%	0.88
9%	1.27
12%	1.53
15%	1.53
18%	1.49

Slope of the Lift Curve

It is often useful to know the slope of the lift curve (obtained by plotting C_L versus angle of attack) in the flight regime. When the slope is indicated with reference to the angle of attack in degrees, it is designated by the lower-case letter a. When indicated in radians, the lower-case letter

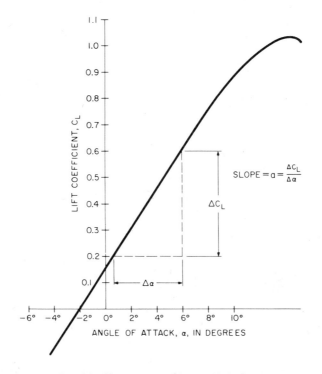

Fig. 35 Determining Slope of Lift Curve

Slope of the Lift Curve (cont'd)

m is used. One unit is easily converted to the other by means of the following equation,

$$m = 57.3a$$

Typical values for a series of airfoils of the same family, but varying in thickness ratio, are:

Thickness Ratio	Slope (m) for Aspect Ratio = 6
6%	4.28
9%	4.28
12%	4.32
15%	4.24
18%	4.20

The slope of the lift curve is equal to the tangent of the curve in the straight portion of the curve, or expressed mathematically,

$$a = \tan \frac{\Delta C_L}{\Delta a}$$

which is read, "a is equal to the tangent whose value is the ratio of delta C_L to delta alpha." The term delta is used to indicate small increments. Figure 35 is a graphical representation of the method used to determine the slope.

Angle of Attack Variation

The slope of the lift curve, as shown in Fig. 35, varies with the aspect ratio. The lift coefficients do not change, but the angles of attack do. The change may be calculated from the following equation:

$$a_{AR} = a_o + \frac{57.3 C_L}{\pi AR}$$

where a_{AR} is the angle of attack, in degrees, to be determined
a_o is the angle of attack, in degrees, for the value of C_L at infinite aspect ratio, since the ratio $57.3 C_L / \pi AR = 0$ when $AR = $ infinity
C_L is the independent variable
AR is the aspect ratio for which a_{AR} is desired

If the available data define an aspect ratio other than infinite, and the data for a different aspect ratio are desired, it becomes necessary to calculate a_o first.

Angle of Attack Variation (cont'd)

Thus,

$$a_0 = a_{AR_1} - \frac{57.3 C_L}{\pi AR_1}$$

where AR_1 is the aspect ratio for the known data

Having found a, the a_{AR_2} corresponding to the new aspect ratio can be calculated:

$$a_{AR_2} = a_0 + \frac{57.3 C_L}{\pi AR_2}$$

$$= a_{AR_1} - \frac{57.3 C_L}{\pi AR_1} + \frac{57.3 C_L}{\pi AR_2}$$

$$= a_{AR_1} + \frac{57.3 C_L}{\pi} \left[\frac{1}{AR_2} - \frac{1}{AR_1} \right]$$

Since the calculations are repetitive, it is convenient to set them up in tabular form to evaluate the values of the above equation, which can be further simplified to read

$$a_2 = a_1 + 18.24 C_L \left[\frac{1}{AR_2} - \frac{1}{AR_1} \right]$$

where $57.3/\pi = 18.24$

Example: A wing has the following characteristics: Aspect ratio = 6, C_L = 0.25, at an angle of attack of 4 degrees. What would be the angle of attack for the same lift coefficient if the aspect ratio of a wing with the same airfoil were increased to 10?

Solution: From the data, $a_1 = 4°$, $C_L = 0.25$, $AR_1 = 6$, $AR_2 = 10$. Substituting in the equation,

$$a_2 = 4 + 18.24 \, (0.25) \left[\frac{1}{10} - \frac{1}{6} \right]$$

$$= 4 + 18.24 \, (0.25) \, (-0.0667)$$

$$= 4 - 0.304 = 3.7 \text{ degrees}$$

A table similar to the one on page 59 would be useful.

ADVANCED AIRFOIL AERODYNAMICS

COMPUTATION OF ANGLE OF ATTACK									
C_L	−1.0	−0.8	−0.6	−0.4	−0.2	0.0	0.2	0.4	0.6
a_1									
$\Delta a = 18.24 K C_L$									
$a_2 = a_1 + \Delta a$									

Note that $\frac{57.3}{\pi} = 18.24$ and $K = \frac{1}{AR_2} - \frac{1}{AR_1}$ which becomes a constant for any given sets of aspect ratios. Also, when the airfoil characteristics are known for aspect ratio infinity, $K = 1/AR_2$.

The Drag Coefficient

This also varies with angle of attack and with changes in aspect ratio. A typical graph of the drag coefficient versus angle of attack is shown in Fig. 36.

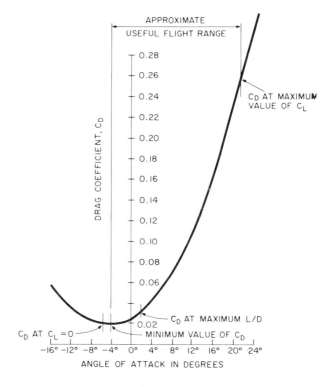

Fig. 36 Plot of C_D vs Angle of Attack

The Drag Coefficient (cont'd)

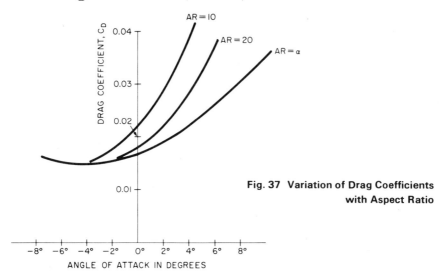

Fig. 37 Variation of Drag Coefficients with Aspect Ratio

The variations of the drag coefficient with aspect ratio are shown in Fig. 37. The drag coefficients for infinite aspect ratio are called *profile drag coefficients* or *section* drag coefficients.

The drag of an airfoil is composed of two elements, one known as the profile drag, usually designated by D_o, and the *induced drag* (D_i) which is a function of the lift characteristics and the aspect ratio of the wing of the airfoil. It is more convenient to deal with the coefficient. Thus, if

$$D = D_o + D_i$$

then

$$C_D = C_{D_o} + C_{D_i}$$

Induced Drag

The induced drag coefficient can be found from the relationship

$$C_{D_i} = C_L^2/AR$$

The profile drag coefficient remains constant for an airfoil. In other words, it is independent of the lift coefficient and of change in aspect ratio. If the data available for the drag coefficient are for one aspect ratio, other than infinite, and the drag coefficient for a different aspect

Induced Drag (cont'd)

ratio is desired, it becomes necessary to calculate C_{D_0} for the aspect ratio having available data, or

$$C_{D_0} = C_{D_{AR_1}} - C_L^2/\pi AR_1$$

where AR_1 is the aspect ratio for the known data

Having found the value for C_{D_0}, then $C_{D_{AR_2}}$ corresponding to the new aspect ratio can be calculated as follows:

$$C_{D_{AR_2}} = C_{D_0} + C_L^2/\pi AR_2$$

$$= C_{D_{AR_1}} - C_L^2/\pi AR_1 + C_L^2/\pi AR_2$$

$$= C_{D_{AR_1}} + \frac{C_L^2}{\pi}\left[\frac{1}{AR_2} - \frac{1}{AR_1}\right]$$

The induced drag coefficient decreases with increase in aspect ratio. In other words, a high aspect ratio is desired (consistent with structural considerations) if the drag coefficient is to be as low as possible. (Sailplanes use high aspect ratios to reduce drag.)

Example: A wing of aspect ratio 10 has a drag coefficient of 0.001, at a lift coefficient of 0.020. What would the drag coefficient be if the same airfoil were employed for a wing with an aspect ratio of 6?

Solution: The applicable formula reads

$$C_{D_2} = C_{D_1} - C_L^2/\pi AR_1 + C_L^2/\pi AR_2$$

$$= C_{D_1} + \frac{C_L^2}{\pi}\left[\frac{1}{AR_2} - \frac{1}{AR_1}\right] = C_{D_1} + K_1 C_L^2$$

where $(1/\pi)(1/AR_2 - 1/AR_1)$ reduces to a constant and may be designated by a suitable letter such as K_1

For the problem stated above,

$$C_{D_1} = 0.001^2 \quad C_L = 0.020^2 \quad AR_1 = 10^2 \quad AR_2 = 6$$

Therefore,

$$K_1 = \frac{1}{\pi}\left(\frac{1}{6} - \frac{1}{10}\right) = \frac{0.667}{\pi} = 0.212$$

Induced Drag (cont'd)

so that,

$$C_{D_2} = 0.001 + 0.212C_L^2 = 0.001 + 0.0000848$$

or no change to all intents and purposes. However, the change is appreciable for larger values of the lift coefficient and cannot be neglected.

Since these calculations are repetitive, it is convenient to set up a table like the one shown. This table can be combined with that for determining the angle of attack.

COMPUTATION OF DRAG COEFFICIENT											
C_L	−1.0	−0.8	−0.6	−0.4	−0.2	0.0	0.2	0.4	0.6	0.8	etc.
C_{D_1}											
C_L^2											
$KC_L^2 = \Delta C_D$											
C_{D_i}											
$C_{D_2} = C_{D_1} + \Delta C_D$											

If the profile drag of the airfoil is desired, the following equation is used:

$$C_{D_o} = C_{D_{AR}} - C_L^2/\pi AR$$

Thus,

$$C_{D_o} = 0.001 - \frac{(0.020)^2}{6\pi} = 0.001 - 0.0000212$$

or, the C_{D_o} value at the lift coefficient 0.020 is, to all intents and purposes, the same as the drag coefficient. However, as the lift coefficients approach 1.0 and above, the difference is appreciable.

ADVANCED AIRFOIL AERODYNAMICS

Aspect Ratio Corrections

Experimental data for the aerodynamic characteristics are usually given for a specific aspect ratio. *Aspect ratio* corrections are made by recalculating the angle of attack and the drag coefficient.

There may be other corrections due to interferences with the fuselage, or engine nacelles, wing tip shapes, and the like, but these corrections are highly empirical and depend upon a considerable body of experimental data for proper interpretation.

Aerodynamic Twist

A wing is said to have *aerodynamic twist* when the angle of incidence of successive chord-wise sections outboard from the aircraft plane of symmetry is different. For example, in a simplified case, the chord of the airfoil at the wing tip could be at an angle 3 degrees greater than the chord of the airfoil at the root of the wing. The angle of incidence of each chord between the root and the tip would vary at a linear rate, from zero at the root to 3 degrees at the tip.

Aerodynamic twist improves spanwise lift distribution and thus results in better flight characteristics. The lift and drag coefficients for such a wing would be an average for all the airfoils represented from root to tip, referred to the angle of attack of the mean geometric chord.

Lift-Over-Drag Ratio

One measure used in determining the choice of an airfoil for a particular design is the lift-over-drag ratio (L/D). Referring to the lift and drag equations, it is easily seen that

$$L/D = C_L/C_D$$

These values are calculated for each angle of attack. A typical graphical representation is shown in Fig. 38.

Comparing the figures for lift and drag shows that the maximum value on the L/D curve does not occur at the minimum drag coefficient angle, nor at the same angle as the maximum lift coefficient.

Center of Pressure

The algebraic summation of the pressures over the entire surface of the airfoil, or wing, produces a resultant force that has a definite mag-

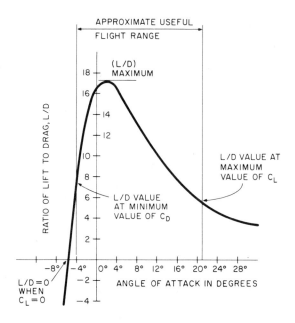

Fig. 38 Typical L/D Curve for Airfoil

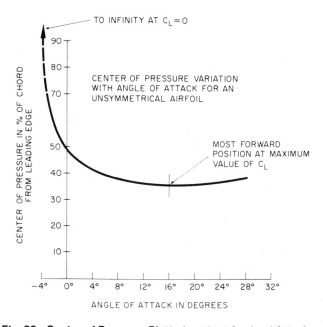

Fig. 39 Center of Pressure Plotted against Angle of Attack

Center of Pressure (cont'd)

nitude and direction. It also must have a definite point of application known as the *center of pressure* (usually abbreviated c.p.).

The location of the center of pressure aft of the leading edge of the airfoil is usually expressed as a percentage of the chord. For subsonic airfoils in common use today, the most forward position of the center of pressure is located at between 25 and 30 percent of the chord. The variation of the location of the center of pressure along the chord line is shown graphically in Fig. 39.

Moment Coefficients

As stated above, the pressure distribution over an airfoil can be represented by a single equivalent force, or resultant, and the point at which it acts is called the center of pressure. It is convenient to refer forces to other reference points. In doing so, a moment—or a couple as it is sometimes called—is introduced so that the same effects are obtained. As an illustration, the three systems shown in Fig. 40 are equivalent as far as the effect on any point between b and c is concerned. The moment (M) could be represented by an equivalent couple a distance d apart, and equal to $(P/d)(L - x)$. The couple $(P/d)(L - x)$ could be shown either at an angle or vertically, as long as the two vectors were parallel.

Let's consider point c in each of the diagrams shown in Fig. 40a. Taking moments about c in Fig. 40a, the algebraic relationship is

$$M_c = PL$$

For that shown in Fig. 40b,

$$M_c = Px + M_b$$
$$= Px + P(L - x)$$
$$= Px + PL - Px = PL$$

For that shown in Fig. 40c,

$$M_c = Px + \frac{P}{d}(L - x)d$$
$$= Px + PL - Px = PL$$

Since the moments about point c are the same for all three configurations, the three systems are equivalent. Thus, in moving a force from one point to another, a moment such as M_D has to be added to obtain an equivalent system. It is the second configuration, Fig. 40b, that becomes of interest

Moment Coefficients (cont'd)

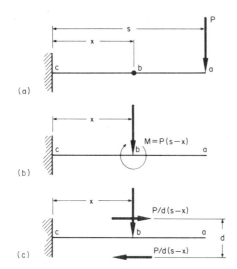

Fig. 40 Equivalent Force and Moment Systems

when the aerodynamic forces are referred to the aerodynamic center (which corresponds to point b in Fig. 40) from the center of pressure (which corresponds to point a in Fig. 40).

Instead of dealing with moments, which are expressed in inch-pounds or foot-pounds, it is convenient to deal with coefficients.

Aerodynamic Center

At the point on the airfoil known as the *aerodynamic center* (a.c.), the moment coefficient remains constant throughout the useful range of aerodynamic characteristics. It happens that the aerodynamic center for the standard airfoil falls between 23 and 27 percent of the chord. It may also fall slightly above or below the assumed chord line. The moment equation is of the same form as before,

$$M = CSq\, C_{M\,a.c.}$$

where M = the moment, in foot-pounds, if C is expressed in feet; in inch-pounds, if C is expressed in inches
C = the chord of the airfoil, in feet or inches. (The former is more common. For a wing, the reference chord is the mean aerodynamic chord.)
S = the wing area, in square feet
q = dynamic pressure = $1/2\,\rho v^2$ where v is expressed in feet per second
$C_{M\,a.c.}$ = moment coefficient about the aerodynamic center

Aerodynamic Center (cont'd)

Example: What is the pitching moment of a wing with an area of 400 square feet and a speed of 250 miles per hour under standard sea-level conditions? The mean geometric chord of the wing is 8 feet. The moment coefficient is

$$C_{M_a} = -0.008$$

Solution: The applicable formula is

$$M = qC_{M_a} CS$$

where C_{M_a} = -0.008
C = 8.0 ft
S = 400 sq. ft
q = $1/2 \rho v = 1/2 (0.002378)(250 \times 1.467)^2$

Substituting these values,

$$M = 1/2 (0.002378)(250 \times 1.467)^2 (-0.008)(8)(400)$$
$$= -4100 \text{ foot-pounds}$$

The pitching moment is negative, indicating that it is a diving moment tending to pitch the nose of the wing down. This moment would have to be counteracted by an opposite moment produced by the tail surfaces.

While the calculations were calculated in terms of foot-pounds, the structural engineer would be interested in obtaining these pitching moments in inch-pounds. Multiplying the above answer by 12 would give the desired result.

Section Characteristics

While early data often called for an aspect ratio of 6, recent works give aerodynamic characteristics in terms of infinite aspect ratio; hence, the data are called *section characteristics.* The corresponding lift and drag coefficients carry the lower case letter as a subscript rather than the capital letter (C_l rather than C_L; C_d rather than C_D).

Recapitulation

The aerodynamic characteristics of an airfoil are usually shown on a graph such as Fig. 41. It is convenient to plot the values of C_L, C_D, c.p.,

Recapitulation (cont'd)

$C_{M_{a.c.}}$, $C_{M_{C/4}}$, and L/D as ordinates. The angle of attack, α, is usually used as the abscissa, although C_L is the primary independent variable, of which all the other characteristics are functions. (For convenience, the center of pressure ordinates are inverted to avoid cluttering the lower half of the chart.)

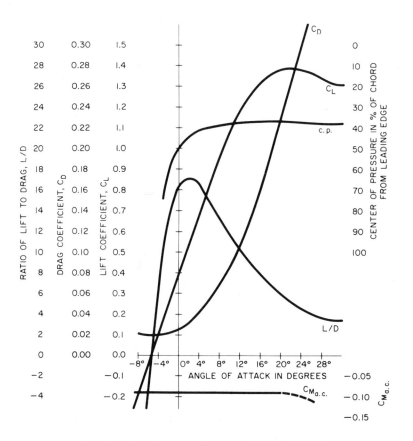

Fig. 41 Aerodynamic Characteristics of Airfoil

Review Questions

1. A rectangular wing employs an airfoil for which $C_L = 0.840$ and $C_D = 0.0598$, for an angle of attack = 8 degrees. The aspect ratio for the wings is 6.
   (a) What is the angle of attack and the drag coefficient, for the same lift coefficient for an aspect ratio of 9; of 12?
   (b) What conclusions may be drawn as to the effect of increasing the aspect ratio upon the drag coefficient? Upon the angle of attack?
   (c) What is the L/D value for each of the three aspect ratios? What conclusions may be drawn as to the effect of increasing the aspect ratio upon the value of the L/D ratio?
2. Determine the slope of the lift curve for an airfoil with the following specifications:

Angle of Attack	C_L
2°	0.50
4°	0.65
6°	0.80
8°	0.95
10°	1.10
12°	1.25

3. What is the slope, in terms of radians, for the airfoil characteristics given in question 2?

NON-AIRFOIL DRAG

Parasite Drag

The aerodynamic characteristics of the airfoil have been discussed in some detail. All the other parts of the airplane have aerodynamic characteristics of their own. Their lift characteristics are usually of minor importance. Their drag characteristics are of great interest; the flight performance of the entire aircraft depends upon the sum of the drag of the component parts.

The drag of the various elements of the airplane is known as *parasite resistance*.

When an airplane is designed, the designer has to calculate performance even before the design gets off the drawing board. The information he needs may come from various sources. Much may be obtained from previous wind tunnel tests on individual parts or part combinations.

Parasite resistance may refer to:

1. Unit length, as for struts of the landing gear.
2. The unit itself, as a wheel or powerplant nacelle.
3. **The maximum cross-sectional area, as for a fuselage, or a powerplant nacelle.**

Such data may be given in terms of a speed—one foot per second or 100 feet per second, or any other speed that the original experimenter found convenient. The variation of these drags with angle of attack is small. The profile of a wheel would not vary with the angle of attack of the airplane, for example.

Therefore, to obtain a useful basis for subsequent performance calculations for a given design, it is necessary to refer them to a common basis, preferably calculating the drags of all parts of the design for the same speed. When this has been done, a new drag coefficient can be based on the wing area by using the basic drag equation,

$$D_p = qSC_{D_p} = 1/2\, v^2 \rho S C_{D_p}$$

where D_p = the sum of all the parasite resistance, in pounds
C_{D_p} = a nondimensional parasite resistance coefficient
S = wing area, in square feet
v = the speed for which the drags of the individual components were obtained

Parasite Drag (cont'd)

Of course, the individual parasite drag coefficients can be calculated and totaled to give one value.

Interference Drag

There is still another parasitic drag, caused by the juxtaposition of the various components of the airplane. There is mutual interference in the airflow between the wing and fuselage; between the powerplant nacelles and the wing or fuselage, or both; between the landing gear and the unit to which it is attached. This is known as *interference drag*. It is difficult to appraise. It is obviously less for a "clean airplane" or for one with the landing gear retracted than for one with the gear exposed.

Example: For preliminary calculations, a designer has allowed the following parasite resistance values for the components indicated:

Tail surfaces:	0.40 pounds per square foot of surface area, at 100 mph
Fuselage:	7.0 pounds per square foot of frontal area, at 100 mph
Engine nacelles:	as for the fuselage
Landing gear:	assumed retracted in the wing

The wing area of his proposed design is 400 square feet, and the total tail surface area is 140 square feet. The frontal area of the fuselage and engine nacelles is 12 square feet. What is the parasite resistance coefficient referred to the wing area?

Solution:
The drag of the tail surfaces at 100 mph is 140 (0.40) = 56 pounds.
The drag of the fuselage and the engine nacelles at 100 mph is 12(7.0) = 84 pounds.
The total parasite resistance is 140 pounds.
The applicable formula is

$$D_p = (1/2) \rho v^2 S C_{D_p}$$

Solving for C_{D_p}:

$$C_{D_p} = D_p / (1/2) \rho v^2 S$$

Interference Drag (cont'd)

where D_p = 140 pounds
ρ = 0.002378 slugs per cubic foot
v = 100 x 1.467 feet per second
S = 400 sq. ft

C_{D_p} = (140)/ (1/2) (0.002378) (100 x 1.467)^2 (400)
= 0.0137

Review Questions

1. What are profile drag, induced drag, parasite drag, and interference drag?
2. How is the profile drag of an airfoil determined?
3. How is the individual drag of an airfoil determined?
4. How is parasite resistance data usually presented?

LIFT INCREASE DEVICES

Various methods of increasing the lifting capacity of a wing have been explored. The maximum obtainable lift coefficient has been of the greatest interest. The reason becomes apparent when the lift equation is examined.

For Minimum Speeds

One of the design criteria for an airplane is the landing speed. This is close to the stalling speed of the airplane, and is obtained at an angle at which the lift coefficient is maximum. If we rearrange the terms of the lift equation,

$$L = (1/2)\rho v^2 C_L S$$

we see that

$$v = \sqrt{\frac{L}{(1/2)\rho C_L S}}$$

but since the maximum lift of an airplane in horizontal flight is its weight, W, it becomes evident that

$$V_{min} = \sqrt{\frac{W}{(1/2)\rho C_{L_{max}}}}$$

where V_{min} stands for the minimum speed
$C_{L_{max}}$ stands for the maximum value of C_L

Early experimenters found that the easiest way to obtain higher values of $C_{L\,max}$ was to change the thickness ratio or to increase the median camber of the airfoil. Both these methods have practical limitations. For example, at low values of the lift coefficient, C_L (occurring at low angles of attack), there is an increase in the drag coefficient, C_D. This increase is a penalty that is always present, so other solutions had to be found.

For Minimum Speeds (cont'd)

A possible solution would seem to be to have two airfoils, one available throughout most of the flying range, and another available for minimum speed only. Such a solution is not practical.

Median Camber Change

The median camber of an airfoil can be changed arbitrarily with a flap, as illustrated in Fig. 42. A number of designs for flaps and their operation have been invented, varying from a simple flap to a multiple set of flaps nesting in the trailing edge. (The latter, when extended, resemble venetian blinds.) Schematic diagrams of commonly used flaps are shown in Fig. 43.

Types of Flaps

The simple flap has its most common application as ailerons, elevators, rudders, and as *tabs* on control surfaces. It was originally developed when it was discovered that deflection of such flaps at the wing tips produced a greater rolling moment than the older method of warping the wings. Similarly, it was found that a flapped surface for the horizontal and vertical tail surfaces was more effective than changing the angle of attack of a plain, unflapped control surface. The plain flap is easy to construct and does not require any intricate operation mechanism.

The Fowler flap is the one most commonly used for transport aircraft. This flap moves toward the rear as it is depressed downward. The result is not only an increase of the median camber, but also an increase of the effective wing area.

Maximum Lift Coefficient for Flaps

The change in the lift coefficients obtained with various types of flaps is shown in Fig. 44. Not only is the lift coefficient increased, but the angle at which zero lift occurs is shifted to larger negative angles. The slopes of the lift curves are not materially changed.

The table on page 76 presents the main characteristics of the lift increase devices shown in Fig. 43. The maximum lift coefficients in the table are for a full-span flap. In actual practice, lift increase devices are not used at the wing tips, since the ailerons are there. The maximum lift coefficient of a wing with a lift increase device will depend upon the relative proportion of the span used by the device.

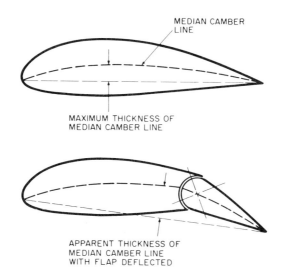

Fig. 42 How Median Camber Line Can Be Changed

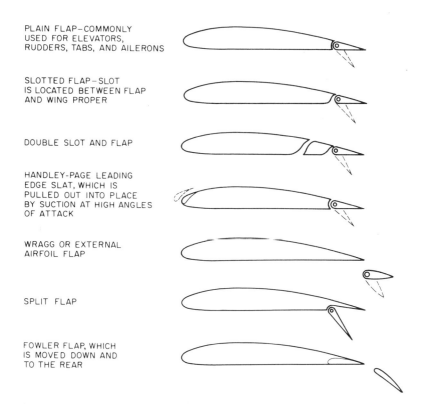

Fig. 43 Methods of Increasing Maximum Lift Coefficient

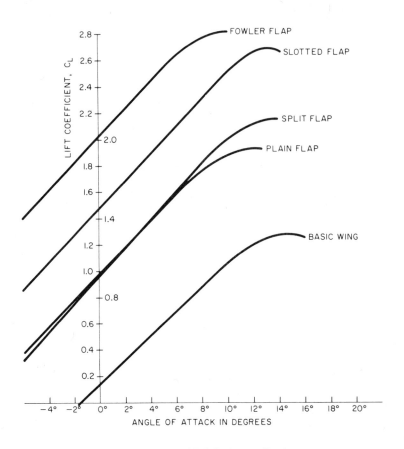

Fig. 44 Effect of Lift Increase Devices

CHARACTERISTICS OF LIFT INCREASE DEVICES

Lift Increase Device	Ratio of Flap Chord to Wing Chord	Deflection of Flap	$C_{L\,max}$	a
Basic Airfoil	0.30	45°	1.29	15°
Plain Flap	0.30	45°	1.95	12°
Slotted Flap	0.30	45°	2.65	13°
Handley-Page Slat	—	—	1.84	28°
Handley-Page Slat and Flap	0.30	45°	2.18	19°
Handley-Page Slat and Fowler Flap	0.40	45°	3.36	16°
Split Flap	0.30	45°	2.25	14°
Fowler Flap	0.30	40°	2.82	10°

Maximum Lift Coefficient for Flaps (cont'd)

It seems not to matter whether the base airfoil is thick or thin as far as attaining the higher lift coefficient is concerned. This is fortunate; airfoils with low thickness ratio are best for high speed performance. The fact that the maximum lift coefficient of the thin airfoils is not high does not matter, because lift increase devices can make up for the lack when needed for lowering the landing speed.

Review Questions

1. What lift increase devices are normally used on wings?
2. What is their purpose?
3. A rectangular wing with a span of 40 feet and a chord of 8 feet has a flap type of lift increase device extending over 50 percent of the span. The lift coefficient of the wing without the flap is 1.2, while that portion having the flap deflected attains 1.8 at the same time. What is the average lift coefficient of the wing?

EXPERIMENTAL METHODS

Before an airplane is constructed, the designer obtains all the experimental data pertaining to his design that he can find. These data may have been obtained previously in his company's laboratory, or in publications available from the U. S. Government Printing Office in Washington. These publications are based on the investigations of NASA. Foreign countries have comparable publications.

After the aerodynamicist has assembled his data and worked out his design, he still has to make small-scale tests on models; it would be prohibitively expensive and perhaps even disastrous to make full-scale tests on untried designs.

The Wind Tunnel

The most efficient tool used in experimental investigations is the wing tunnel. Only the basic principles of the various wind tunnels will be described here. The wind tunnel is what its name implies. A simple subsonic design is shown in Fig. 45.

The size of the wind tunnel is given in terms of the dimensions of the cross section of the working section where the investigations take place.

The Measuring System

A model is suspended from a measuring apparatus to measure lift, drag, and pitching moments. A schematic diagram of such a suspension system is shown in Fig. 46.

The tests must be run without any air speed to obtain the tare readings for each angle of attack. These tare readings are subtracted from the lift and drag readings to obtain net figures. From the net figures, an additional deduction is made to account for the drag of the suspension itself. Any suitable weighing system may be hooked up to each of the suspension wires.

Method of Operation

The tunnel is operated at a constant wind speed. The forces are measured. From these measurements, the coefficients may be calculated from the usual equations.

EXPERIMENTAL METHODS

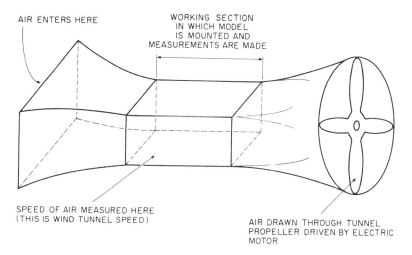

Fig. 45 Sketch of Open Return Wind Tunnel

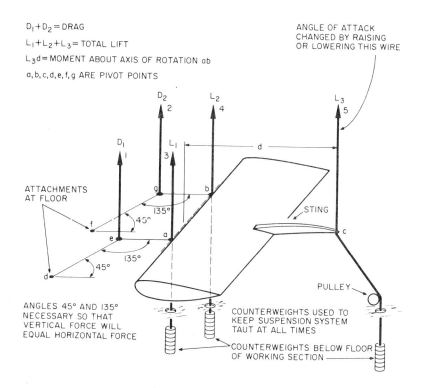

Fig. 46 Setup for Measuring Aerodynamic Forces

Method of Operation (cont'd)

For example, from the equation,

$$L = (1/2)\rho v^2 C_L S$$

the known quantities are:

L = Lift, measured by the weighing balance, in pounds
ρ = mass density of the air where the tests took place. The density of the air depends upon temperature, barometric pressure, and humidity.
v = speed at which the wind tunnel is operated, in feet per second
S = the wing area of the model, in square feet

C_L can now be calculated since all other quantities in the equation are known. Lift is obtained for successive angle of attack values, which can be changed by rotating the model about one of the axes of suspension, as indicated in Fig. 46.

Models

Models of the full-scale prototype are tested in the wind tunnel. The size of the model depends upon the size of the wind tunnel to be used in the investigations. The small models are usually made of a high grade of mahogany, which is less susceptible to humidity variations than other available materials. Not every characteristic of the full-size object needs to be duplicated; for example, structural ridges, door knobs, and the like are not simulated.

The models may be from 1/20 to 1/40 full size, so the question may arise as to the validity of the results found. Correction factors are, of course, applied. Such correction factors sometimes can be obtained only by comparing full-scale values with experimental model values. However, one scale factor, the Reynolds number, is of special interest.

Reynolds Number

An English physicist, Osborne Reynolds, studied fluid flow in pipes, but the conclusions he drew have found application to airflow around objects and have paved the way for understanding the need for aerodynamic similarity. He discovered that, in a pipe, the fluid flow changed from a smooth, or laminar flow, to turbulent flow, as the speed was in-

Reynolds Number (cont'd)

creased. He finally derived the expression now known as the Reynolds number, abbreviated RN:

$$RN = \frac{\rho vR}{\mu}$$

where ρ = the mass density of the fluid
 v = the velocity of the airflow, usually expressed in feet per second
 μ = coefficient of viscosity of the fluid, in pound-seconds per square foot
 R = a characteristic dimension of the object under test. In Reynolds' day, the R was the radius of the pipe in feet. For aircraft, it could be the chord of the wing, or the diameter of the fuselage (if the fuselage were tested alone)

The ratio μ/ρ is known as the kinematic viscosity, v, which, under standard conditions, is 1.576×10^4 at 59° F (15° C).

Aerodynamic Similarity

The aerodynamic similarity of the model would be the same as that of the full-scale prototype provided that:

(RN) model = (RN) full-scale

that is, if the two Reynolds numbers are the same. This is not always possible to achieve for a small model, but the tests on the model may be run at a series of speeds to determine if any radical changes in coefficient values take place. These tests are conducted to determine "scale effect."

Example: Wind tunnel tests are to be conducted on a model of an airplane whose full-scale prototype will have a mean geometric chord (MGC) of 8 feet. The highest speed expected is 250 miles per hour. The wind tunnel available for testing the scaled model limits the size of the model so that it is only one-fortieth of the full scale.

What is the Reynolds number of the full-scale model? At what speed would the wind tunnel have to be operated to get the same Reynolds number?

Solution:

v = 250 x 1.467 feet per second
μ/ρ = 1.576×10^{-4}
R = the characteristic dimension, chosen here to be the chord (R = C = 8 feet)

82 FUNDAMENTALS OF AIRCRAFT FLIGHT

Aerodynamic Similarity (cont'd)

Since, the Reynolds number is RN = $\rho vR/\mu$, where v is in feet per second, for the full-scale prototype,

$$RN = \frac{(250 \times 1.467)(8)}{1.576 \times 10^{-4}} = 936 \times 10^4$$

To obtain the same Reynolds number for the one-fortieth scale model, the available data is

RN = 936×10^4
μ/ρ = 1.576×10^{-4}
R = C = 8/40 since the model is 1/40 full size.

Therefore,

$$RN = 936 \times 10^4 = \frac{(V \times 1.467)(8)}{(1.576 \times 10^{-4})(40)}$$

Solving for V,

$$V = \frac{(936 \times 10^4)(1.576 \times 10^{-4})(40)}{8 \times 1.467}$$

It would have been simpler to equate the unevaluated quantity for the full-scale Reynolds number to obtain

$$\frac{(250 \times 1.467)(8)}{1.576 \times 10^{-4}} = \frac{(V \times 1.467)(8)}{(1.576 \times 10^{-4})(40)}$$

which quickly simplifies to

$$250 = \frac{V}{40}$$

or

$$V = 10{,}000 \text{ miles per hour}$$

Obviously it is not practical to operate the small wind tunnel at this wind speed, which is a value corresponding to about Mach 15. By testing the model at a number of speeds, even if below the 250 miles of the full-scale prototype, meaningful data can, nevertheless, be obtained, since the scale effect is of a minor order.

Review Questions

1. An airplane is flying at 250 miles per hour at 10,000 feet. The mean geometric chord of the wing is 10 feet. At what Reynolds number is the airplane operating?
2. A model of the airplane mentioned in question 1 is made to a scale of $1'' = 20''$. It is tested in a wind tunnel operating under atmospheric conditions. At what speed would the wind tunnel have to operate so that the Reynolds number for both the full-size airplane and its model will be the same?
3. The model in Question 2 was tested in the wind tunnel at the tunnel speed of 60 miles per hour. The lift and drag coefficients at three angles of attack were:

Angle of Attack	C_L	C_D
2°	0.50	0.045
4°	0.65	0.052
6°	0.80	0.066

What were the lifts and drags measured by the wind tunnel balances?

MOTIONS OF THE AIRPLANE

Since an airplane is operating in a three-dimensional field in the air that surrounds it, its motions may be in any direction and about any of its three body axes. These motions are:

Linear Motion

The most important and significant motion is the speed forward parallel to the X-axis of the airplane. Whether this speed is achieved in horizontal flight, climbing flight, or in a dive is immaterial.

An airplane could have a sideward velocity when experiencing a gust from the side, or when any improper turn has been made. Such motion is a *skid*.

The airplane usually loses altitude when in a skid, because the lateral axis of the airplane wing is no longer horizontal. The motion then is a *side-slip*.

Curvilinear Motion

The airplane can turn about any of its three body axes (Fig. 47). Its motion about the Y-axis in the X-Z plane is known as *pitching*. When

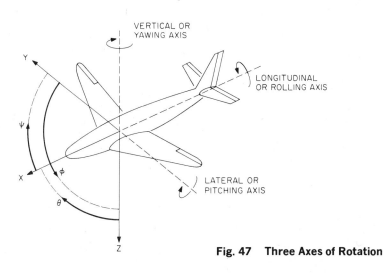

Fig. 47 Three Axes of Rotation

Curvilinear Motion (cont'd)

the airplane nose pitches upward, the motion is considered positive, and the airplane is said to be *stalling* (even though the angular increase is relatively small and the wing is not at a stalling angle, nor is the power plant stalling). (There should be no confusion as to what is meant when the word is used in the proper context.)

When the nose of the airplane pitches down, the airplane is said to be *diving,* although, again, the situation may not be as critical as the term *dive* might indicate.

Motion about the Z-axis in the X-Y plane is known as *yawing.* The airplane is said to be yawed about the Z-axis.

Motion of the airplane about the X-axis in the Y-Z plane is *rolling.*

While the airplane may pitch without any accompanying rolling or yawing, it is rare that either rolling or yawing is not accompanied by the other motions.

Angles

The angles through which the turning motions take place have their own designations, with Greek-letter symbols:
1. The angle of pitch, theta (θ).
2. The angle of roll, phi (ϕ). When an airplane is put into a limited roll to achieve a specified angle, the angle may be called the angle of bank. This is the angle that the airplane has to be placed in to make a coordinated turn, and would thus be accompanied by a yawing motion.
3. The angle of yaw, psi (ψ).

Review Questions

1. What motions take place about:
   (a) the vertical axis?
   (b) the lateral axis?
   (c) the longitudinal axis?
2. What motions take place along:
   (a) the longitudinal axis?
   (b) the vertical axis?
   (c) the lateral axis?
3. What are the three angles normally considered when rotation takes place?

RECTILINEAR FLIGHT

An airplane is in *rectilinear flight* when it is flying at a steady speed, in the vertical plane—that is, without changes of direction in the horizontal plane. The three basic modes of flight in the vertical plane are *horizontal flight, climbing flight,* and *gliding flight.* The *vertical dive* is a phase of gliding flight.

Vertical Forces in Horizontal Flight

Consider the airplane flying horizontally in still air, in unaccelerated flight, at an angle of attack a. The line of action of the relative wind is along the flight path and opposite to the direction of flight. The forces acting on the airplane are depicted in Fig. 48.

The airplane is in *static equilibrium,* since there is no acceleration. Thus, all the forces balance each other. Accordingly, one may say that the summation of the vertical forces must equal zero. Symbolically, this may be written

$$\Sigma V = 0$$

Since it is often convenient to use X and Y axes, this same equation may be written

$$\Sigma Y = 0$$

The convention used is that forces acting upward are considered positive, and those downward, negative. For the forces shown in Fig. 48, this equation may be expanded as follows:

$$\Sigma Y = L_w + L_T + T \sin(a + i) - W = 0$$

In examining this equation, one may make some assumptions that will simplify without invalidating the conclusions.

The angle of attack a, is the angle included between the flight path and the reference chord, usually the Mean Geometric Chord (MGC) of the wing.

L_T represents the lift, in pounds, of the tail surfaces. It is small compared with the lift of the wing, and can be ignored in the equation above.

$T \sin(a + i)$ is the vertical component of the thrust furnished by the power plant. Angle i is the angle included between the line of action

Vertical Forces in Horizontal Flight (cont'd)

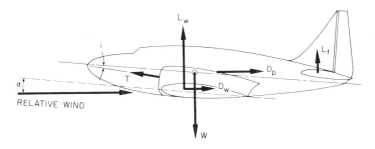

Fig. 48 Forces Acting in Horizontal Flight

of the thrust and the reference chord, MGC. The maximum value of $a + i$ may be as much as 20 or more degrees; however, smaller angles are being considered here. The sin of about 6 degrees or less is below 0.10, while the thrust is of the order of magnitude of the total drag, which is also about one-tenth of the lift. The contribution of the thrust, $T \sin(a + i)$, to lifting the aircraft may therefore be neglected.

W is the gross weight of the airplane. The vector for the weight acts vertically through the center of gravity of the airplane, where the whole mass of the airplane can be assumed to be concentrated.

L_w represents the lift of the wings. For convenience, the subscript will be dropped in subsequent discussion. The lift contribution of other components of the aircraft such as the fuselage, engine nacelles. etc., are small and need not be considered in this discussion.

With the assumptions made so far, the equation

$$\Sigma Y = 0$$

reduces to

$$L - W = 0$$

or, more conveniently

$$L = W$$

Some Conclusions

The first conclusion one may draw is that, in horizontal flight, the lift must be equal to the weight of the aircraft. This would be realized intuitively, but here is the proof.

Some Conclusions (cont'd)

Earlier it was established that

$$L = (1/2)\rho v^2 C_L S$$

By equating the expressions for L and solving for v, we obtain

$$v = \sqrt{\frac{2W}{\rho C_L S}}$$

This is an interesting equation, for it can lead to several conclusions, especially when considered in conjunction with the known characteristics of the airfoil for which a typical graph is shown in Fig. 49. Examination of the last equation reveals that the speed (v) of the airplane varies:

> *Directly* with the wing loading. Once the aircraft has been built, the wing area (S) remains constant. The weight (W) varies in flight as the fuel is consumed. However, if two aircraft are compared, the one with the greater wingloading would have to fly faster at the same value of C_L than the lighter, to have sufficient lift to maintain the same altitude.

> *Inversely,* with the value of lift coefficient. The greater the lift coefficient, the lower the speed required to maintain horizontal flight. The characteristic lift curve in Fig. 49 indicates that there is a maximum value for the lift coefficient C_L. It should be obvious that the lowest speed obtainable in horizontal flight is at the maximum lift coefficient. This speed is known as the stalling speed of the aircraft. For all practical purposes, the stalling speed may be considered the landing speed. (Note: The stalling speed of the aircraft is that determined by the value of the maximum lift coefficient. It is not to be confused with stalling of the power plant.)

> *Inversely,* as the air density, ρ. The aircraft will have to fly faster at higher altitudes to maintain flight for the same angle of attack, than at sea level.

One may draw some other conclusions by looking at the characteristic lift curve:

> For each lift coefficient, there is only one angle of attack.

> For $C_L = 0$, the speed would have to be infinite, implying infinite drag and infinite power. Hence, there is a maximum limit to the attainable speed.

Some Conclusions (cont'd)

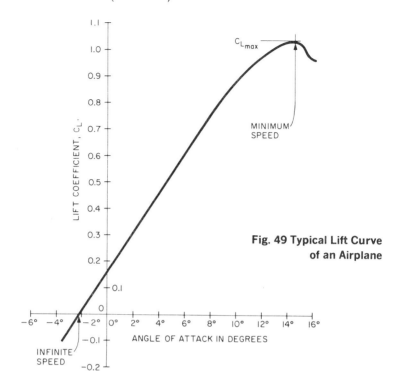

Fig. 49 Typical Lift Curve of an Airplane

For negative values of C_L, there would be no solution, because the square root of a negative number is imaginary. These lift coefficients could be realized only if the airplane were flying upside down, when the signs of the lift coefficient would be reversed.

The flight regime lies between $C_{L\ max}$ and some value of C_L above zero.

If the minimum (or landing) speed and the gross weight of the aircraft are fixed, the aircraft with the highest value of $C_{L\ max}$ would have the least wing area. For this reason, the maximum lift coefficients of a series of airfoils are compared. However, if some of the available lift-increase devices (such as flaps, slots, and combinations thereof) are used, the maximum lift coefficient of the basic airfoil becomes less important.

What is not apparent from either the equation or the lift curve is that normal flight is not possible beyond the angle of the maximum lift coefficient. The stall of the wing is to be avoided since the airplane cannot then be controlled about any of its three axes.

Some Conclusions (cont'd)

Example: An airplane has a wing area of 250 feet and a gross weight of 3000 pounds. It is flying in horizontal flight at standard sea-level conditions at an angle of attack of 4 degrees, at which angle the lift coefficient $C_L = 0.28$. What must be the speed of the airplane to maintain horizontal flight?

Solution: The pertinent formula is

$$v = \sqrt{\frac{2W}{\rho C_L S}}$$

where $W = 3000$ pounds
$\rho = 0.002378$ slugs per cubic foot
$C_L = 0.28$
$S = 250$ square feet

Thus,

$$v = \sqrt{\frac{2(3000)}{(0.002378)(0.28)(250)}}$$

$$= \sqrt{36,300}$$

$$= 190.5 \text{ feet per second}$$

$$= 130 \text{ miles per hour}$$

Variation of Speed with Altitude

The speed at any other altitude can be calculated from the fact that at sea level

$$v_0 = \sqrt{\frac{2W}{\rho_0 C_L S}}$$

where the subscript zero designates sea-level conditions

At any altitude (h),

$$v_h = \sqrt{\frac{2W}{\rho_h C_L S}}$$

Variation of Speed with Altitude (cont'd)

Dividing the second equation by the first,

$$\frac{v_h}{v_0} = \sqrt{\frac{\rho_0}{\rho_h}}$$

so that

$$v_h = v_0 \sqrt{\frac{\rho_0}{\rho_h}}$$

Example: An airplane with a wing area of 300 square feet, and a gross weight of 4500 pounds is flying at an angle of attack of 8 degrees, at which the lift coefficient C_L is 0.84. How does its speed in horizontal flight, at 10,000 feet, compare with that at sea level?

Solution: Reference to the table on ICAO Standard Atmosphere (see page 8) indicates that ρ at sea level is 0.002377, and is 0.001655 at 10,000 feet altitude. From the statement of the problem, it is known that W = 4500 pounds, S = 300 square feet, and C_L = 0.84. The applicable formula for sea level is

$$v_0 = \sqrt{\frac{2W}{\rho_0 C_L S}}$$

$$= \sqrt{\frac{2\,(4500)}{(0.002377)\,(0.84)\,(300)}}$$

= **122.67 feet per second, or 83.7 miles per hour**

The applicable formula for the 10,000-foot altitude is

$$v_h = \sqrt{\frac{2W}{\rho_h C_L S}}$$

$$= \frac{2\,(4500)}{(0.001655)\,(0.84)\,(300)}$$

= **146.79 feet per second, or 100.5 miles per hour**

The speed at 10,000 feet could have been found from the relationship once the speed, v_0, at sea level was determined.

$$v_h = v_0 \sqrt{\frac{0.002377}{0.001655}}$$

$$= v_0 \sqrt{1.435} = 1.197\, v_0$$

Variation of Speed with Altitude (cont'd)

Substituting for v_0,

$$v_h = 1.197\,(122.67)$$
$$= 146.8 \text{ feet per second}$$

or

$$v_h = 1.197\,(83.7)$$
$$= 100 \text{ miles per hour}$$

(The slight discrepancies which appear are due to the fact that both slide rule and longhand calculations were made.)

Horizontal Forces in Horizontal Flight

In this case, the summation of forces in the horizontal direction must equal zero:

$$\Sigma H = 0$$

or using the X-Y axis system,

$$\Sigma X = 0$$

Conventionally, forces acting to the right are considered positive, and those toward the left, negative.

Referring to Fig. 48,

$$\Sigma X = -T\cos(a+i) + D_p + D_w = 0$$

where $T\cos(a+i)$ is the horizontal component of the thrust in pounds. It may be assumed that the value of $\cos(a+i)$ is close to 1 (since even assuming the cosine of 25 degrees to be 1 would cause an error slightly less than 10 percent)

D_w is the drag, in pounds, of the wing

D_p is the drag, in pounds, of all components of the airplane other than the wing. It is called *parasite drag*. This drag can be referred to wing area of the aircraft so that one may write

$$D_p = (1/2)\rho v^2 C_{D_p} S$$

The appropriate equation for the drag of the wing is

$$D_w = (1/2)\rho v^2 C_D S$$

Horizontal Forces in Horizontal Flight (cont'd)

With these assumptions and simplifications, one may write

$$-T + D_p + D_w = 0$$

or

$$T = (1/2)\rho v^2 S(C_{D_p} + C_D)$$

This equation indicates that the thrust must be equal to the total drag to maintain horizontal flight. Of course, this conclusion could have been arrived at intuitively.

It is useful to substitute for v^2 in the equation for thrust its equivalent $2W/\rho C_L S$ so that

$$T = \left[(1/2)S(C_{D_p} + C_D)\right]\left[\frac{2W}{\rho C_L S}\right]$$

which results in

$$T = \frac{W(C_D + C_{D_p}^{tot})}{C_L} = \frac{W}{(L/D)_{tot}}$$

where

$$(L/D)_{tot} = \frac{C_L}{C_D + C_{D_p}}$$

A typical curve for the lift-to-drag ratio for an airplane is very similar to that for the wing alone. The two are compared in Fig. 50.

Variation of Thrust

The equation indicates that the thrust required to maintain horizontal flight varies:

Directly as the weight. In other words, the greater the weight of the aircraft, the more thrust is required.

Inversely as the lift-to-drag ratio L/D. Therefore, the greater the ratio, the lower the thrust required to maintain horizontal flight. It is, therefore, important to make the aircraft as clean as possible, and to choose an airfoil from a series that has the best L/D ratio.

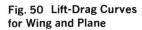

Fig. 50 Lift-Drag Curves for Wing and Plane

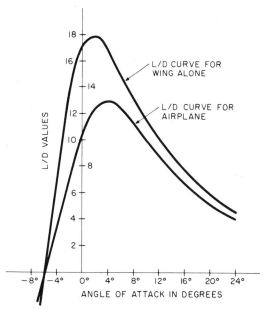

Variation of Thrust (cont'd)

Independently of the altitude. While a change in air density brings about a change in speed, the value of $1/2\,\rho v^2$ remains the same. Another way of looking at it is that the value of L/D is independent of the air density. Similarly, the gross weight is also independent; thus, the thrust needed to overcome the drag for a given airplane at any given angle of attack is independent of altitude.

Note that the L/D ratio is zero when C_L is zero, and that, therefore, the thrust required would be infinite just as the speed would have to be infinite. Thrust for an airplane is furnished by either a jet engine or a propeller-engine combination. The principles of these thrust-producing mechanisms are discussed in other books of this series.

Example 1: An airplane with a gross weight of 6000 pounds is flying at an angle of attack of 4 degrees in horizontal flight. The characteristic curves for the wing and the airplane are shown in Fig. 50. What is the drag of the wing and of the complete airplane? What is the thrust required to maintain the airplane in horizontal flight?

Solution: Reading the chart as carefully as possible, the value of the lift-to-drag-ratio for the wing alone at 4 degrees is 16.8, and for the air-

Variation of Thrust (cont'd)

plane, 12.8. Since, in horizontal flight, the lift of the airplane equals its weight and is assumed to be borne by the wing (or $L = W$), then,

$$D = W / \frac{L}{D} = \frac{6000}{16.8} = 357 \text{ pounds, the drag of the wing.}$$

For the complete airplane,

$$D = W / \frac{L}{D} = \frac{6000}{12.8} = 469 \text{ pounds, the total drag of the airplane.}$$

In horizontal flight, the thrust must equal the total drag, hence the thrust that the power plant must furnish is

$$T = D = 469 \text{ pounds}$$

Example 2: What would be the drag for the wing and the airplane at the same angle of attack of 4 degrees if the airplane were in horizontal flight at 10,000 feet? At 20,000 feet?

Solution: The drag and weight relationships do not change. The applicable equation is the same for all altitudes, or $D = W/(L/D)$. Hence, the drag is the same. Also, the thrust required would remain the same.

Power Required

Thrust is related to power, since work is done when thrust is exerted by the power plant of the airplane flying at a steady speed. Hence, multiplying the thrust (T) in pounds, by v, the speed in feet per second, represents work done in terms of foot-pounds per second. Since one horsepower is defined as work equal to 550 foot-pounds per second, it follows that $Tv/550$ = horse-power required, or simply,

$$P = \frac{Tv}{550}$$

where P stands for power required

It has already been established that in horizontal flight,

$$T = D_{tot} = (1/2) \rho v^2 S \left(C_D + C_{D_p} \right)$$

Power Required (cont'd)

Substituting for T in the previous equation,

$$P = \frac{v}{550}\left[(1/2)\rho v^2 S\left(C_D + C_{D_p}\right)\right] = \frac{\rho v^3 S}{11{,}000}\left(C_D + C_{D_p}\right)$$

By substituting for v, the equivalent expression $\sqrt{2W/\rho C_L S}$ and simplifying, the following equation is obtained:

$$550\,P = \left[\frac{C_{D_{tot}}}{C_L^{3/2}}\right]\left[\sqrt{\frac{2}{\rho}}\right]\left[\frac{W^{3/2}}{S^{1/2}}\right]$$

$$= \left[\frac{1}{C_L(L/D)_{tot}}\right]\left[\sqrt{\frac{2}{\rho}}\right]\left[\frac{W}{S}\right]^{1/2}[W]$$

Several conclusions may be drawn from this equation. It is apparent that the power required varies:

> *Inversely* as the square root of the air density. Accordingly, as the airplane reaches higher altitudes, more power is required to maintain horizontal flight at the same angle of attack than at sea level. Recall that it was pointed out that the airplane has to fly faster at altitude than at sea level at the same angle of attack. Internal combustion engines must also be supplied with an air compressor to deliver the necessary power.
> *Inversely* as the L/D ratio. Hence, the higher this ratio, the less power required.
> *Directly* as the square root of the wing loading.

Example: An airplane with a gross weight of 8000 pounds is in horizontal flight at 10,000 feet. Its wing area is 500 square feet and the angle of attack in flight is 8 degrees. The parasite resistance coefficient, referred to the wing area, is $C_{D_p} = 0.028$, which is assumed to have little variation with angle of attack. The wing characteristics are shown in Fig. 41. Determine the following for the angle of attack indicated:

(a) speed in horizontal flight at 10,000 feet.
(b) the L/D of the airplane.
(c) the total drag of the airplane.
(d) the thrust required to fly at this speed.
(e) the horsepower required to fly at this speed.

Power Required (cont'd)

Solution: From the statement of the problem: W = 8000 pounds, S = 500 square feet, α = 8 degrees, altitude H = 10,000 feet, C_{Dp} = 0.028. From the table listing the ICAO Standard Atmosphere (see page 8) characteristics: ρ_h = 0.001655. From Fig. 41, interpolating as carefully as possible: C_L = 0.975 and C_D = 0.068.

(a) The speed is determined from the equation

$$v = \sqrt{\frac{2W}{\rho C_L S}}$$

$$= \sqrt{\frac{2(8000)}{(0.001655)(0.975)(500)}}$$

$$= 148.6 \text{ feet per second or } 101.2 \text{ mph}$$

(b) The L/D of the airplane is obtained from the formula

$$L/D = \frac{C_L}{C_{D_w} + C_{D_p}}$$

Since C_L = 0.975, C_{D_w} = 0.068, and C_{D_p} = 0.028, then,

$$C_{D_{tot}} = C_{D_w} + C_{D_p} = 0.068 + 0.028 = 0.096$$

and so,

$$L/D = \frac{0.975}{0.096} = 10.15$$

(c) The total drag of the airplane at this altitude and angle of attack in horizontal flight is

$$D = W / \left(\frac{L}{D}\right) = \frac{8000}{10.15} = 788 \text{ pounds}$$

This drag is independent of the altitude.

(d) The thrust required to fly at this speed, at 10,000 feet, is determined from the relationship that thrust equals drag; hence,

$$T = D = 788 \text{ pounds}$$

Power Required (cont'd)

(e) The horsepower required to maintain horizontal flight at this altitude and angle of attack can be calculated from the formula

$$P = \frac{Tv}{550} = \frac{788\,(148.6)}{550} = 213 \text{ horsepower}$$

It could also have been calculated from the formula

$$P = \frac{1}{550}\left[\frac{1}{C_L\,(L/D)_{tot}}\right]\left[\sqrt{\frac{2}{\rho}}\right]\left[\frac{W}{S}\right]^{1/2}[W]$$

$$= \frac{1}{550}\left[\frac{1}{0.975\,(10.15)}\right]\left[\sqrt{\frac{2}{0.001655}}\right]\left[\frac{8000}{500}\right]^{1/2}[8000]$$

(The student will find it interesting to calculate the same quantities for the airplane for an angle of attack of 4 degrees.)

Variation of Power with Altitude

The power at any altitude, once the power at sea level has been calculated, can be determined from the fact that at sea level,

$$550\,P_o = \left[\frac{C_{D_{tot}}}{C_L^{3/2}}\right]\left[\sqrt{\frac{2}{\rho_o}}\right]\left[\frac{W^{3/2}}{S^{1/2}}\right]$$

and at any altitude, h

$$550\,P_h = \left[\frac{C_{D_{tot}}}{C_L^{3/2}}\right]\left[\sqrt{\frac{2}{\rho_h}}\right]\left[\frac{W^{3/2}}{S^{1/2}}\right]$$

where the subscripts o and h indicate sea level and altitude h conditions

Dividing one equation by the other,

$$\frac{P_h}{P_o} = \sqrt{\frac{\rho_o}{\rho_h}} \quad \text{and} \quad P_h = P_o\sqrt{\frac{\rho_o}{\rho_h}}$$

The variations of horsepower required with change in altitude is shown graphically in Fig. 51.

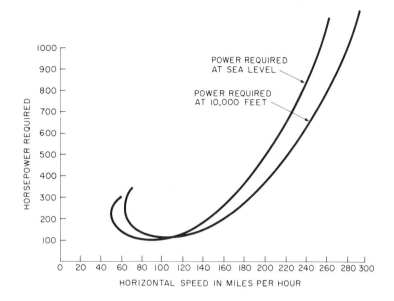

Fig. 51   Typical Curves of Power Required

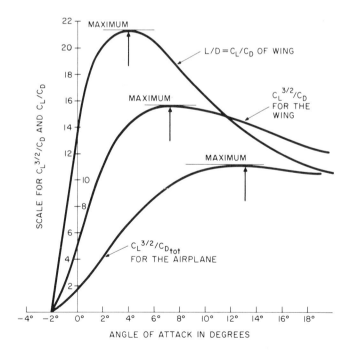

Fig. 52   Power Coefficients for Wing and Plane

Power Coefficient

Since the power required is proportional to

$$\frac{C_{D_{tot}}}{C_L^{3/2}} = \frac{C_D + C_{D_p}}{C_L^{3/2}}$$

the ratio of $C_L^{3/2}/C_D$ is often called a "power coefficient." Since $C_D/C_L^{3/2}$ should be a minimum if P is to be a minimum, it follows that $C_L^{3/2}/C_D$ has to be a maximum.

Values of $C_L^{3/2}/C_D$ have been calculated for a typical airfoil and plotted against the corresponding angles of attack as shown in Fig. 52.

For purposes of comparison, the curve for the L/D values for the wing is also shown. (Note that minimum power is required at a relatively large angle of attack for the complete airplane.)

Example: At 8 degrees angle of attack, a wing has a lift coefficient (C_L) of 0.975 and a drag coefficient (C_D) of 0.068. The parasite resistance coefficient of the airplane referred to the wing area is $C_{D_p} = 0.028$. What is the power coefficient for the wing? What is the power coefficient for the airplane?

Solution:

$$C_L = 0.975 \qquad C_D = 0.068 \qquad C_{D_p} = 0.028$$

$$\text{Power coefficient} = C_L^{3/2}/C_D$$

The power coefficient (let's call it P_C for ease in referring to it below) for the wing is then

$$P_C = (0.975)^{3/2}/0.068 = (0.962)/0.068 = 14.15$$

For the airplane, it would be

$$P_C = (C_L^{3/2})(C_D + C_{D_p}) = (0.975)^{3/2}/(0.068 + 0.028) = 10.00$$

Review Questions

For the problems below, the following information is furnished: Gross weight = 2500 pounds, wing area = 240 square feet, aspect ratio of the wing = 6. The lift and drag coefficients of the wing for three different angles of attack are:

Angle of Attack	C_L	C_D
2°	0.50	0.025
4°	0.65	0.032
6°	0.80	0.046

The parasite drag coefficient of airplane, referred to the wing area, may be taken as 0.020.

Calculate the following:
1. The airplane speed necessary to maintain horizontal flight at each angle of attack at sea level.
2. The wing drag, the parasite drag, and the total drag at each of the three angles of attack.
3. The airplane speed at each of the angles of attack at 5000, 10,000, and 15,000 feet altitudes.
4. The L/D of the airplane at each of the three angles of attack.
5. The power coefficients for each angle of attack.

ENCOUNTERING A GUST

Effect of a Gust

When an airplane in horizontal flight encounters gusty conditions, the forces acting on it change. The designer has to anticipate these changes to assure a safe design. (The action of the gust has led to the erroneous conception of the "air pocket," a term used by early pilots who did not understand the behavior of gusty air at higher altitudes.)

To understand the effect of a gust, assume that an airplane is flying at a speed of v feet per second, in a steady horizontal flight. It encounters a momentary upward gust of wind with a speed of u feet per second. The increase in the angle of attack, since it is small, may be designated $\triangle \alpha$.

Figure 53 shows the change in angle of attack caused by an upward gust as well as by a downward gust. The tangent of this angle is

$$\tan \triangle \alpha = u/v$$

When $\triangle \alpha$ is expressed in radians, the value of the tangent of a small angle can be taken as being equal to the angle without serious error, or

$$\tan \triangle \alpha_R = \triangle \alpha_R$$

where R is used to indicate that the angle is expressed in radians

It follows, therefore, that

$$\triangle \alpha_R = u/v$$

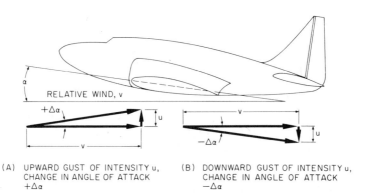

(A) UPWARD GUST OF INTENSITY u, CHANGE IN ANGLE OF ATTACK $+\triangle \alpha$

(B) DOWNWARD GUST OF INTENSITY u, CHANGE IN ANGLE OF ATTACK $-\triangle \alpha$

Fig. 53 Change in Attack Angle due to Gust

Change in Lift Coefficient

Along with the incremental change in the angle of attack ($\triangle a$) there is an incremental change in the lift coefficient, or $\triangle C_L$. It is useful to express $\triangle a_R$ in terms of $\triangle C_L$. This can be done as indicated in Fig. 54. Note that the slope of a curve at any point is the tangent drawn to the curve at that point. Since the lift curve is a straight line through the greater portion of its useful range, the tangent coincides with the straight-line "curve." If this slope is called m, then

$$m = \triangle C_L / \triangle a_R$$

Solving for $\triangle a_R$,

$$\triangle a_R = \triangle C_L / m$$

But $\triangle a_R$ also equals u/v from which it follows that

$$\triangle C_L = mu/v$$

Before the airplane encountered the gust, the lift coefficient was C_L; after encountering the gust, it was $C_L + \triangle C_L$.

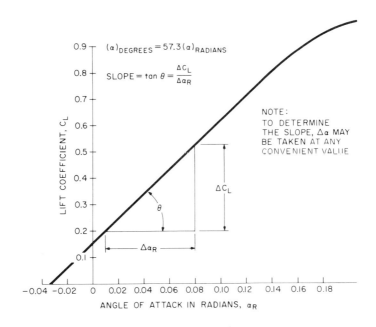

Fig. 54 Finding the Slope of the Lift Curve

Change in Lift Coefficient (cont'd)

The velocity of the relative wind striking the airplane before entering the gust was v. The resultant velocity after encountering the gust would be practically the same. Even for gust velocities of 20 percent of the forward velocity, which is exceedingly high, the resultant speed would be only about 10 percent higher than the speed of the airplane before it encountered the gust.

The lift in steady horizontal flight before the gust was

$$L = (1/2)\, \rho v^2 C_L S = W$$

Effect of Gust on Lift

The lift upon encountering the gust would be

$$L' = (1/2)\, \rho v^2 (C_L + \Delta C_L) S$$

Dividing the second equation by the first

$$\frac{L'}{L} = \frac{L'}{W} - \frac{C_L + \Delta C_L}{C_L} = 1 + \frac{\Delta C_L}{C_L}$$

Substituting for ΔC_L its value mu/v, and for C_L its value $W/[(1/2)\,\rho S v^2]$ the equation reduces to

$$\frac{L'}{W} = 1 + \frac{mu/v}{W\,[(1/2)\,\rho S v^2]} = 1 + \frac{\rho muv}{2(W/S)}$$

Load Factor

The ratio of L'/W is called the load factor, and usually designated n. Referring to the equation just obtained, it could be written as

$$n = 1 + \Delta n$$

where Δn is the incremental load factor

The meaning of the equation can be explained thus: Every item of the airplane would increase in apparent weight by a factor of Δn. A person's weight would be n times his weight, having increased by Δn times his weight, at the time the airplane encountered the gust.

Load Factor (cont'd)

If the gust were vertically downward, the incremental load factor would be negative or

$$n = 1 - \Delta n = 1 - \frac{\rho m u v}{2(W/S)}$$

Accordingly, the same person's weight would be lighter.

The load factor is sometimes referred to as the number of g's encountered in a gust. Here, g is the gravitational constant.

(The same effect as a gust is encountered in an elevator. As the elevator comes to a stop, there is an increase of one's foot pressure on the elevator floor. As the elevator starts suddenly downward, there is a decrease in the foot pressure.)

This derivation is based upon a sharp-edged gust, and the assumption that the airplane is rather rigid. An empirical factor or constant is usually introduced, and since the gust may be up or down, the minus sign is also indicated so that

$$\frac{L'}{W} = n = 1 \pm \frac{K\rho m u (1.467 V)}{2(W/S)} = 1 \pm \frac{K m u V}{575(W/S)}$$

where K varies from 0.2 to 1.2 for wing loading W/S from zero to 50
 m is the slope per radian of the lift curve. For aspect ratios of 6, the value for m lies between 4.07 and 4.34
 ±u is the gust velocity in feet per second. Values of 15 to 30 feet per second are assumed for commercial aircraft
 V is the flight speed, in miles per hour, upon entering the gust. To convert to feet per second, multiply V by 5280/3600 = 1.467

For sea-level conditions, the constants are

$$\frac{1.467}{2} \rho = \frac{1.467(0.002378)}{2} = \frac{1}{575}$$

Example: An airplane with a gross weight of 8000 pounds and a wing area of 500 square feet is flying at sea level at an angle of attack of 8 degrees for which the lift coefficient (C_L) of the wing is 0.975. The slope, m, of the lift curve of the wing is 4.0. While flying, it encounters a vertical upward gust of 30 feet per second. What is the load factor?

Load Factor (cont'd)

Solution: The formula for the load factor is

$$\frac{L'}{W} = n = 1 + \frac{KmuV}{575(W/S)}$$

The available data is: m = 4.0, u = 30, W = 8000 pounds, S = 500 square feet; hence, the wing loading W/S = 8000/500 = 16.

From Fig. 55, the value of the gust factor K for a wing loading of 16 can be found. It is 1.01, as closely as one can read the chart.

The speed V has to be found. It is obtainable from the equation

$$V = \frac{v}{1.467} = \frac{1}{1.46}\sqrt{\frac{2W}{\rho S C_L}}$$

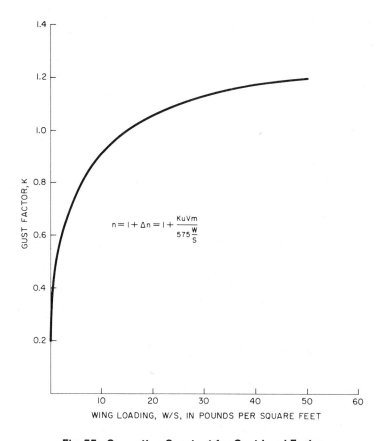

Fig. 55 Correction Constant for Gust Load Factor

ENCOUNTERING A GUST

Load Factor (cont'd)

Substituting the numerical quantities,

$$V = \frac{1}{1.467} \sqrt{\frac{2\,(8000)}{(0.002378)\,(500)\,(0.975)}}$$

$$= 80.2 \text{ miles per hour}$$

The load factor may now be calculated from the formula

$$n = 1 + \frac{KmuV}{575\,(W/S)}$$

$$= 1 + \frac{(1.01)\,(4.0)\,(30)\,(80.2)}{575\,(16)}$$

$$= 1 + 1.055 = 2.055$$

In the airplane, every item would weigh slightly more than twice as much. A person with his entire weight on his seat would press down on his seat with a pressure 2.055 times his weight.

If the gust were vertically down instead of up, the load factor would be

$$n = 1 - \frac{KmuV}{575\,(W/S)} = 1 - 1.055 = -0.055$$

In other words, for this particular case, every item would be practically weightless. It would weigh only 0.055 times its weight. A 100-pound object would weigh, momentarily, 5.5 pounds.

Review Questions

1. What is the weight increment for a person weighing 200 pounds on an airplane whose wing loading is 15 pounds per square foot flying at 250 miles per hour upon entering an upward gust of 30 feet per second? The slope of the lift curve is m = 4.00.
2. What would be the load factor if a downward gust of 30 feet per second were encountered?

CLIMBING FLIGHT

Forces Acting on the Airplane

In climbing, as in gliding, the flight path is inclined to the horizontal. The forces acting on the airplane are the same as before, but are at a different angle to the gravitational axis. The lines of action of the lift and drag forces are used as convenient coordinate axes. The lift forces are perpendicular to the relative wind; the drag forces are parallel to it. Figure 56 represents the force configuration for an airplane in the climb.

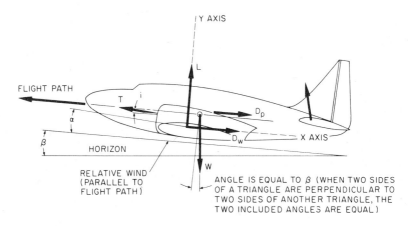

Fig. 56 Forces Acting on Airplane in Climb

Summation of Forces

Again, since the airplane is in equilibrium, the summation of forces must equal zero. Hence,

$$\Sigma Y = 0$$

or

$$L_w + L_T + T_a \sin(\alpha + i) - W \cos \beta = 0$$

Likewise,

$$\Sigma X = 0$$

Summation of Forces (cont'd)

or

$$-T_a \cos(a + i) + D_p + D_w + W \sin \beta = 0$$

The same assumptions can be made as before: that L_T and $\sin(a + i)$ are small quantities that can be neglected, and that the value of $\cos(a + i)$ is close to 1. Transferring terms from one side of the equation to the other,

$$L_w = W \cos \beta$$

and

$$T_a - D_{tot} = W \sin \beta$$

where $D_{tot} = D_p = D_w$

Speed along Climb Path

Substituting $(1/2)\rho v^2 C_L S$ for W and solving for v (but designating it v_c to show that it is the speed along the climb path), the equation becomes

$$v_c = \sqrt{\frac{W \cos \beta}{(1/2) \rho S C_L}}$$

From this, one concludes that since $\cos \beta$ is less than 1 for any angle of climb, the speed along the climb path is somewhat slower than in horizontal flight at the same angle of attack. However, for all practical purposes, $\cos \beta$ may be considered equal to 1. (The square root of the value for $\cos 36°$ would be 0.899, so that assuming $\cos \beta = 1$ would lead to an error of only slightly more than 10 percent for an angle of climb far larger than usually achieved.)

All the other conclusions drawn for ρ, C_L, and W/S in the case of horizontal flight apply as well in this case.

Example: Compare the speed along a climb path with that in horizontal flight for an airplane of 8000 pounds gross weight, 500 square feet of wing area, climbing at an angle of climb of 30 degrees, at an angle of attack to the flight path of 8 degrees, for which the lift coefficient of the wing is $C_L = 0.0975$. The L/D of the complete airplane is 10.15.

Speed along Climb Path (cont'd)

Solution: For horizontal flight the applicable formula is

$$v = \sqrt{\frac{2W}{\rho S C_L}}$$

And for climbing flight,

$$v_c = \sqrt{\frac{2w \cos \beta}{\rho S C_L}}$$

Since the problem did not state at what altitude the comparison was to be made, it will be assumed that standard sea-level conditions prevail. The data available is then: $\rho = 0.002378$, $C_L = 0.975$, $W = 8000$ pounds, $S = 500$ square feet, $\cos \beta = \cos 30° = 0.866$, and $L/D = 10.15$.

In horizontal flight,

$$v = \sqrt{\frac{2\,(8000)}{(0.002378)\,(500)\,(0.975)}}$$

$$= 117.4 \text{ feet per second or } 80.2 \text{ miles per hour}$$

In climbing flight,

$$v_c = \sqrt{\frac{2\,(8000)\,(0.866)}{(0.002378)\,(500)\,(0.975)}}$$

$$= 110 \text{ feet per second or } 75 \text{ miles per hour}$$

Angle of Climb

The angle of climb (β) can be expressed in terms of the velocity (v_c) along the flight path, and in terms of the rate of climb (a) which is the velocity in the vertical direction. Referring to Fig. 57, it can be seen that

$$\sin \beta = \frac{a}{v_c}$$

or

$$a = v_c \sin \beta$$

Angle of Climb (cont'd)

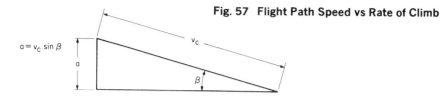

Fig. 57 Flight Path Speed vs Rate of Climb

Since

$$T - D_{tot} = W \sin \beta$$

then

$$T - D_{tot} = \frac{Wa}{v_c}$$

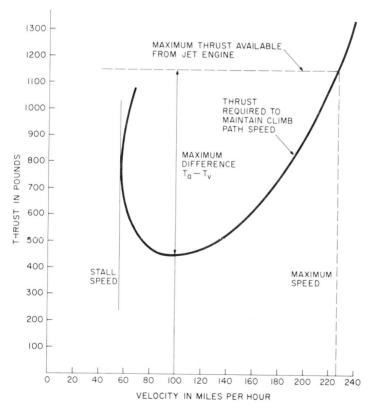

Fig. 58 Thrust vs Velocity

Angle of Climb (cont'd)

The thrust required can be approximated since it can be assumed to be almost equal to that required in horizontal flight. The value of $\cos \beta$ can be assumed to be equal to one without introducing too great an error, so that (L = W) holds and thus, the speed can be calculated.

The drag along the flight path is obtainable once the speed along the path is known.

Typical curves for thrust required for the airplane and thrust available for a jet engine plotted against velocity are shown in Fig. 58.

The difference in thrust $(T_a - T_r)$ is equal to $(T_a - D_{tot})$; hence,

$$T_a - T_r = W \sin \beta$$

or

$$\sin \beta = \frac{T_a - T_r}{W}$$

where T_a = thrust available, in pounds
T_r = thrust, in pounds, required to overcome the drag

Rate of Climb

From the differences in thrust $(T_a - T_r)$ as shown in Fig. 58, the rates of climb can be calculated for

$$T_a = T_r = \frac{Wa}{v_c}$$

Solving for a, and noting that $(T_a - T_r)$ represents the excess thrust, the equation becomes

$$a = \frac{v_c \text{ (excess thrust)}}{W}$$

For each value of excess thrust, the corresponding v_c can be found; hence, the equation can be evaluated.

Power Required

For internal combustion engines operating a propeller, it is more convenient to deal with power calculations.

Power Required (cont'd)

Multiplying both sides of the equation

$$T_a - D_{tot} = Wa/v_c$$

by $v_c/500$ gives

$$\frac{T_a v_c}{550} - \frac{(D_{tot} v_c)}{550} = \frac{Wa}{550}$$

where $T_a v_c/500$ is P_a or horsepower available
$D_{tot} v_c/500$ is P_r or horsepower required

so that $P_a - P_r$ represents the excess horsepower available for the climb. Solving for the rate of climb,

$$a = \frac{550(P_a - P_r)}{W}$$

Typical curves for power versus velocity are shown in Fig. 59.

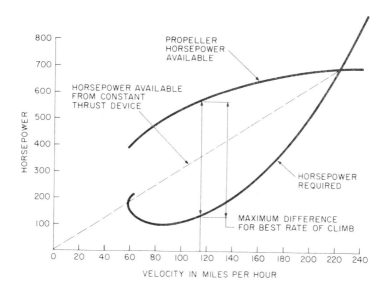

Fig. 59 Horsepower Curves for an Airplane

Power Required (cont'd)

The maximum rate of climb can be calculated from the formula by measuring the maximum difference between the horsepower-available curve and the horsepower-required curve.

The horsepower-available curve for a propeller-engine combination requires extensive calculations for determining propeller efficiencies, as well as knowing the power characteristics of the internal combustion engine. No attempt will be made in this text to cover either the propeller or the powerplant.

Example: The thrust available at an angle of attack of 8 degrees is 30 percent more than required to fly the airplane horizontally. The airplane weighs 8000 pounds and has a wing area of 500 square feet. The lift coefficient of the wing is 0.975 and the L/D of the airplane is 10.15. What is the rate of climb?

Solution: The formula for the rate of climb is expressed in feet per second as

$$a = v_c (T_a - T_r) / W$$

The velocity along the climb path is determined from the formula

$$v_c = \sqrt{\frac{2W \cos \beta}{\rho C_L S}}$$

where β is the angle of climb

The square root of $\cos \beta$ can be assumed to be almost 1 without appreciable error, so the formula becomes

$$v_c = \sqrt{\frac{2W}{\rho C_L S}}$$

The thrust required must equal the drag; hence

$$T_r = W/(L/D)$$

The thrust available (T_a) is 30 percent in excess of that required; hence, $T_a = 1.30\ T_r$. With the information available,

$$v_c = \sqrt{\frac{2\,(8000)}{(0.002378)\,(0.975)\,(500)}} = \textbf{117.4 feet per second}$$

Power Required (cont'd)

And,

$$T_r = D = \frac{W}{L/D} = \frac{8000}{10.15} = 788 \text{ pounds}$$

$$T_a = 1.30 \, T_r = 1.3(788) = 1025 \text{ pounds}$$

The rate of climb

$$a = \frac{v_c (T_a - T_r)}{W} = \frac{117.4 \, (1025 - 788)}{8000} = 2.78 \text{ feet per second}$$

The rate of climb is generally expressed in feet per minute rather than in feet per second, while the speed along the flight path is 117.4 feet per second. Therefore,

Example 2: If instead of a 30 percent excess in thrust, there was a 30 percent excess in horsepower, what would be the rate of climb for the airplane above?

Solution: Where horsepower is involved, the rate of climb is

$$a = \frac{550 \, (P_a - P_r)}{W}$$

This formula requires the calculation of P_r:

$$P_r = \frac{Dv}{550}$$

Since $P_a = 1.30 \, P_r$, the formula for rate of climb reduces to

$$a = \frac{550 \, (1.30 \, P_r - P_r)}{W} = \frac{(0.3)(550)(P_r)}{W}$$

Substituting Dv/500 for P_r, then,

$$a = \frac{(0.3)(550)}{W} \left(\frac{Dv}{550} \right) = \frac{0.3 \, Dv}{W}$$

Power Required (cont'd)

The velocity (v_c) along the climb path was calculated previously:

$$v_c = \sqrt{\frac{2(8000)}{(0.002378)(0.975)(500)}} = 117.4 \text{ feet per second}$$

The drag was calculated previously:

$$D = \frac{W}{L/D} = \frac{8000}{10.15} = 788 \text{ pounds}$$

so that the rate of climb is

$$a = \frac{(0.3)(788)(117.4)}{8000}$$

$$= 3.47 \text{ feet per second or 208 feet per minute}$$

What are the approximate angles of climb in the two cases?
(a) In the case of 30 percent excess thrust, the vertical rise is 2.78 feet per second, while the speed along the flight path is 117.4 feet per second. Therefore,

$$\beta = \sin^{-1} 2.78/117.4 = \sin^{-1} 0.00237 = 1.35°$$

(b) In the case of 30 percent excess power, the vertical rise is 3.47 feet, so that the angle of climb, β, is

$$\beta = \sin^{-1} 3.47/117.4 = \sin^{-1} 0.0295 = 1.7°$$

The substitution of $\sqrt{\cos \beta} = 1$ in the formulas for the climb is therefore justified since

$$\sqrt{\cos 1.7} = \sqrt{0.9996} = 0.9998$$

CLIMBING FLIGHT

Review Questions

For the problems below, the following information is furnished: gross weight = 2500 pounds, wing area = 250 square feet, aspect ratio of the wing = 6. The lift and drag coefficients of the wing for three different angles of attack are as follows:

Angle of Attack	C_L	C_D
2°	0.50	0.025
4°	0.65	0.032
6°	0.80	0.046

The parasite drag coefficient of airplane referred to the wing area may be taken as 0.020. The airplane is climbing.

Calculate:

(a) The rate of climb when the thrust available is 30 percent greater than the thrust required for each of the three angles of attack.
(b) The rate of climb when the horsepower required at each angle of attack is 25 percent greater than the horsepower required.
(c) The angle of climb for each of the cases in (a) and (b).

GLIDING FLIGHT

In gliding flight, the glide path is inclined to the horizontal at an angle beta (β). The forces on the airplane are the same as before: weight, thrust, lift on wing and tail surfaces, parasite and wing drag. The coordinate X-Y axes coincide with lines of action of lift and drag, known as the wind axes. All this is shown in Fig. 60.

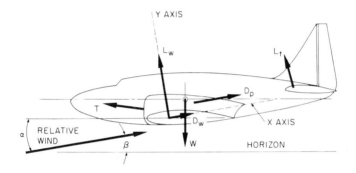

Fig. 60 Forces Acting during a Glide

Summation of Forces

Proceeding as before, by taking the summation of the forces,

$$\Sigma X = 0$$

$$\Sigma X = D_{tot} - T \cos(a + i) - W \sin\beta + L_T = 0$$

As before, it will be assumed that the angle ($a + i$) is small, and that the lift on the tail surfaces is small. Cos ($a + i$) may be assumed to equal one. So the equation reduces to

$$D_{tot} - T - W \sin\beta = 0$$

or

$$W \sin\beta = D_{tot} - T$$

Summation of Forces (cont'd)

Similarly,

$$\Sigma Y = 0$$

$$L - W \cos \beta + T \sin (a + i) = 0$$

but since sin (a + i) may be assumed equal to zero for small angles,

$$L - W \cos \beta = 0$$

or

$$W \cos \beta = L$$

Dividing one equation by the other,

$$\frac{W \cos \beta}{W \sin \beta} = \frac{L}{D_{tot} - T}$$

but since

$$\frac{\cos \beta}{\sin \beta} = \cot \beta$$

then,

$$\cot \beta = \frac{L}{D_{tot} - T}$$

Interpreting the Equations

From this equation, several interesting conclusions may be drawn. When there is no thrust

$$\cot \beta = \frac{L}{D_{tot}}$$

1. The angle of glide is a function of the L/D ratio for the airplane.
2. The best or flattest glide for the airplane occurs when the L/D of the airplane is a maximum.

By examining the lift and L/D curves for a typical airplane (see Figs. 49 and 50), the following additional observations can be made:

Interpreting the Equations (cont'd)

3. There is only one angle of attack for the maximum L/D value.
4. There are two possible angles of attack for L/D values other than the maximum.
5. Since the speed of an airplane is a function of the lift coefficient (C_L), there may be two speeds available for L/D values other than the maximum—with the lower speed occurring at the greater value of the lift coefficient, and at the higher angle of attack.

Effect of Thrust on the Glide

The effect of having some thrust is to increase the value of the cotangent so that the angle of glide at the same angle of attack is flatter when thrust is available.

To learn about the speed of the aircraft in the glide, the equation $L = W \cos \beta$ will be useful. Since

$$L = (1/2) \rho v^2 C_L S,$$

it follows that

$$v_g = \frac{W \cos \beta}{(1/2) \rho C_L S}$$

where v_g represents the speed in the glide

From this equation it can be seen that the velocity along the glide path is less than in the horizontal flight for the same angle of attack. The maximum value of L/D for airplanes varies from about 4 to about 8 in the subsonic ranges. Such L/D ratios would mean angles of glide from about 14 degrees (for L/D = 4) to about 7 degrees (for L/D = 8). The cosine of 14 degrees is 0.97, so that the glide speed would be $\sqrt{0.97}$ or 0.984 times that in horizontal flight, at the same angle of attack.

Sinking Speed

The vertical component of the glide path speed is of particular interest in considering the operation of gliders or non-powered aircraft. (Refer to Fig. 61.) It is called the sinking speed and can be found from the relationship, $v_s = v_g \sin \beta$. This could also be written:

$$v_s = v \sin(\cot^{-1} L/D_{tot})$$

It is read, "sine of the angle for which the cotangent is L/D_{tot}."

Sinking Speed (cont'd)

Fig. 61 Sinking and Gliding Speed Relations

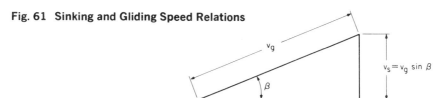

The smaller the value of β, the smaller the value of the sine, and hence, the lower the sinking speed. For a non-powered glider, a very low sinking speed is desirable since the craft can be kept aloft if the vertical component of the wind gusts is as great or greater than the sinking speed.

The sinking speed of an airplane gliding in still air indicates how fast the airplane is losing altitude and reaching the ground.

A graph can be drawn by plotting the glide speeds calculated for a number of glide angles β to obtain both vertical component, or the sinking speed, and the horizontal component. The fourth quadrant is used since the sinking speed is considered negative in value. Such a diagram, as shown in Fig. 62, is called a *hodograph*.

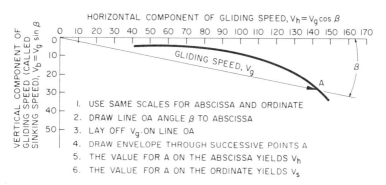

Fig. 62 Velocity Hodograph for Gliding Plane

The Glide Angle

The glide angle is of interest to the pilot who must choose a landing spot for an aircraft in a glide. (Refer to Fig. 63.)

If the aircraft is at an altitude (H) and is gliding at an angle β, it may theoretically land in a circle whose radius (R) bears the relationship to the other quantities,

$$R = H \cot \beta = H(L/D)_{tot}$$

The Glide Angle (cont'd)

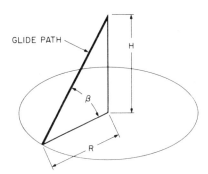

Fig. 63 Landing Circle for Gliding Plane

One may conclude that the larger the value of L/D_{tot}, the larger the circle available for landing.

Landing

The act of landing may be considered in conjunction with gliding flight since the approach to landing is from a glide path. A simplified graphical representation is shown in Fig. 64.

The "flared" landing is achieved when the pilot assumes control by changing the angle of attack of the airplane to achieve a smoother landing. The "dead-stick" landing is achieved by continuing the normal glide until the airplane touches the ground. It usually results in a rougher or harder landing—the flared landing approach is likely to be smoother.

Example 1: An airplane with a gross weight of 8000 pounds and a wing area of 500 square feet is gliding along a glide path without thrust or power at an angle of attack of 8 degrees. The value of C_L is 0.975 and the L/D ratio of the airplane is 10.15. What are:
   (a) The angle of glide?
   (b) The speed along the glide path?
   (c) The sinking speed of the aircraft?
   (d) The landing radius if glide begins at 100 feet altitude?

Solution: The glide angle (β) is determined from the formula

$$\cot \beta = L/D$$

which can be expressed algebraically as

$$\beta = \cot' L/D$$

Landing (cont'd)

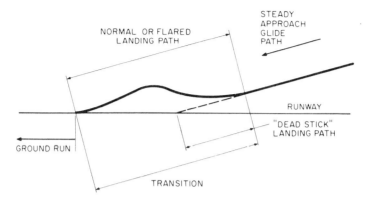

Fig. 64 Stages in the Landing Approach

The gliding speed can be determined from the formula

$$v_g = \sqrt{\frac{2W \cos \beta}{\rho C_L S}}$$

The sinking speed can be determined from the formula

$$v_s = v_g \sin \beta$$

The landing radius can be determined from the formula

$$R = H \cos \beta = H(L/D)$$

The available data is: $W = 8000$ pounds, $S = 500$ square feet, $C_L = 0.975$, L/D of the airplane is 10.15. The glide begins at an altitude of 1000 feet.

(a) The angle of glide (β) calculated from $\cot \beta = L/D = 10.15$. Consulting a trigonometric table, the angle whose cotangent has the value of 10.15 is about 5.6 degrees.

(b) The speed along the glide path is

$$v_g = \sqrt{\frac{2W \cos \beta}{\rho C_L S}} = \sqrt{\frac{2(8000)(\cos 5.6°)}{(0.002378)(0.975)(500)}}$$

$$= \sqrt{\frac{2(8000)(.9952)}{(0.002378)(0.975)(500)}} = 117.3 \text{ feet per second}$$

Landing (cont'd)

The difference in the air density at 1000 feet and sea level is insignificant since it varies from 0.002308 to 0.002378 (a ratio of 0.9804 whose square root is 0.933, so that the variation in the glide speed is less than 7 percent).

(c) The sinking speed is

$$v_s = v_g \sin \beta = v_g \sin 5.6°$$

$$= 117.3 \, (0.0976) = 11.45 \text{ feet per second}$$

This is low sinking speed, and desirable in a calm atmosphere. In a turbulent atmosphere, however, it would make for an uncomfortable descent.

(d) The landing radius is

$$R = \frac{HL}{D}$$

$$= (1000)(10.15) = 10,150 \text{ feet or } 1.925 \text{ miles}$$

Example 2: What effect would a thrust of 100 pounds have upon these same quantities?

Solution: The glide angle (β) is determined from the formula

$$\cot \beta = \frac{L}{D_{tot} - T}$$

The gliding speed can be determined from the formula

$$v_g = \sqrt{\frac{2W \cos \beta}{\rho C_L S}}$$

The sinking speed can be determined from the formula

$$v_s = v_g \sin \beta$$

The landing radius can be determined from the formula

$$R = H\left(\frac{L}{D - T}\right)$$

(a) For determining the angle of glide, it will first be necessary to calculate the total drag.

$$D = \frac{W}{L/D} = \frac{8000}{10.15} = 788 \text{ pounds}$$

Landing (cont'd)

Then

$$\cot \beta = \frac{L}{D_{tot} - T} = \frac{8000}{788 - 100} = 11.61$$

The angle whose cotangent is 11.61 is about 4.9 degrees.

(b) The speed along the glide path is

$$v_g = \sqrt{\frac{2W \cos \beta}{\rho C_L S}} = \sqrt{\frac{2(8000)(\cos 4.9°)}{(0.002378)(0.975)(500)}}$$

$$= \sqrt{\frac{2(8000)(0.9963)}{(0.002378)(0.975)(500)}} = 117.3 \text{ feet per second}$$

(c) The sinking speed is

$$v_s = v_g \sin \beta = 117.3 \sin 4.9$$

$$= 117.3 (0.0854) = 10 \text{ feet per second}$$

(d) The landing radius is

$$R = H\left(\frac{L}{D - T}\right) = 1000 \left(\frac{8000}{788 - 100}\right)$$

$$= 11{,}610 \text{ feet or 2.2 miles}$$

Review Questions

For the problems below, the following information is furnished: Gross weight = 2500 pounds, wing area = 250 square feet, aspect ratio of the wing = 6. The lift and drag coefficients of the wing for three different angles of attack are as follows:

Angle of Attack	C_L	C_D
2°	0.50	0.025
4°	0.65	0.032
6°	0.80	0.046

Review Questions (cont'd)

The parasite drag coefficient of airplane, referred to the wing area, may be taken as 0.020. Assume that the airplane is in a glide.
   (a) What is the angle of glide for the airplane at each of the three angles of attack, assuming that there is no thrust produced by the power plant?
   (b) What is the speed of the airplane along the glide path for each of three angles of attack? Compare these speeds with those previously calculated for horizontal flight.
   (c) What is the angle of glide if a thrust of 100 pounds were available in each of the three cases?
   (d) What is the landing circle if the airplane starts to glide from an altitude of 10,000 feet for each of the cases in (a) and (c) above?

THE DIVE

A vertical dive implies that the flight path of the aircraft is in line with the gravitational axis. The force diagram is shown in Fig. 65.

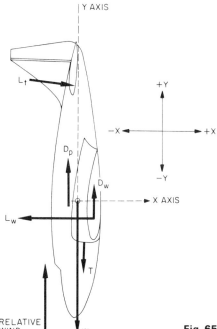

Fig. 65 Forces on an Airplane in the Dive

Summation of Forces

Note that, for convenience as well as custom, the vertical axis has been designated the Y-axis.

$$\Sigma Y = 0$$
$$= -T \cos(\alpha + i) + D_{tot} - W = 0$$

Since $(\alpha + i)$ is small, $\cos(\alpha + i) \approx 1$. Hence $D_{tot} = W + T$. When there is no thrust, $D_{tot} = W$.

$$\Sigma X = 0$$
$$= -L_W - T \sin(\alpha + i) + L_T = 0$$

For small angles, $\sin(\alpha + i) \approx 0$, so that $L_W = L_T$.

Summation of Forces (cont'd)

Since the flight path is vertical, there is no horizontal motion and, thus, the summation of the forces along the horizontal axis must be equal to zero. This implies that the lift on the wing is extremely small, at the very least, and that the magnitude is a function of the lift on the horizontal tail surfaces, whose area is from 10 to 15 percent of the wing area, and, thus, correspondingly small. The equation obtained from ΣY can be combined with the known equation for drag, or

$$D_{tot} = (1/2)\rho v^2 C_{D_{tot}} S$$

and without thrust, T,

$$D = W$$

from $\Sigma Y = 0$. Then, solving for v, and designating it v_d to indicate the speed in the dive:

$$v_d = \sqrt{\frac{W}{(1/2) C_{D_{tot}} S}}$$

Limiting Speed

One concludes that the speed in the dive is maximum when the value of $C_{D_{tot}}$ is a minimum. If a powered dive is considered, the thrust force is added to the effect of W, so a powered dive would attain a greater speed than an unpowered one. With a propeller-equipped airplane, the speed, however, might be such that the propeller would act as a braking rather than a thrusting device, and thus, actually reduce the possible maximum diving speed.

The maximum speed of an aircraft can be obtained in a dive. This speed is important in considering local pressures. Also, the highest load factor can be encountered in pulling out of a dive.

Pulling out of a Dive

It is obvious that, in the dive, the terminal speed is obtained when the value of $C_{D_{tot}}$ is a minimum. Designating this terminal speed v_t, and rewriting the equation,

$$v_t = \sqrt{\frac{W}{(1/2)\rho (C_{D_{tot}})_{min} S}}$$

Pulling out of a Dive (cont'd)

Then, if the airplane were suddenly to be pulled up to an angle of attack corresponding to the maximum lift coefficient, the lift force (L') would be

$$L' = (1/2)\rho v_t^2 C_{L\,max} S$$

Substituting for v_t and simplifying, the equation for L' becomes

$$L' = \frac{C_{L\,max}}{(C_{D\,tot})_{min}} W$$

Dividing each side of the equation by W,

$$\frac{L'}{W} = n = \frac{C_{L\,max}}{(C_{D\,tot})_{min}}$$

The ratio L'/W gives the load factor (n) or the number of g's to which the airplane and the pilot would be subjected if he could pull out of the dive instantly.

The ratio of $C_{L\,max}/C_{D\,min}$ for airfoils alone ranges up to about 25; for an aircraft, the ratio may be in the neighborhood of 12 to 15, so that if it were possible to pull out of a dive abruptly, the lift force might be up to 15 times the weight of the aircraft. However, it is not possible to make a pull-out that abruptly; the pilot doesn't have instant reactions, nor do the controls become effective instantly. The load factor, therefore, would be considerably less than the theoretical value.

It is not practical to design aircraft structures to withstand such high forces, even if the pilot could sustain them, so that the design terminal velocity in the dive is often limited, and must not be exceeded.

Example: An airplane with a gross weight of 8000 pounds and a wing area of 500 square feet is diving at the angle of attack corresponding to the minimum drag coefficient. The parasite resistance coefficient referred to the wing area is $C_{D\,p} = 0.028$. The characteristics of the wing are those shown in Fig. 41.

(a) What is the limiting speed of the airplane in the dive?
(b) If it were possible to pull up out of the dive instantly to the angle of attack at which the wing has its maximum lift coefficient, what would be the potential load factor?

Pulling out of a Dive (cont'd)

Solution: The available data is: $W = 8000$ pounds, $S = 500$ square feet, $C_{D_p} = 0.028$. From Fig. 41, $C_{D_{min}} = 0.020$ and $C_{L_{max}} = 1.38$.

The formula for dive speed is

$$V_d = \sqrt{\frac{2W}{\rho\, C_{D_{tot}}\, S}}$$

The load factor equation is

$$n = \frac{C_{L_{max}}}{C_{D_{tot}}}$$

(a) Solving for the terminal diving speed, the minimum value of the drag coefficient is the determining factor. Therefore,

$$V_d = V_t = \sqrt{\frac{2W}{\rho\, C_{D_{tot}}\, S}}$$

$$= \sqrt{\frac{2\,(8000)}{(0.002378)\,(0.020 + 0.028)\,(500)}}$$

= **5.29.2 feet per second or 360.7 mph**

(b) The potential load factor, if instant pull-out were possible, is obtained from the formula

$$\left(C_{L_{max}}\right) / \left(C_{D_{tot}}\right)_{min} = 1.38/0.048$$
$$= 28.75$$

Review Questions

1. What equation is used to determine the speed in the vertical dive?
2. An airplane of 5000 pounds gross weight and a wing area of 250 square feet is diving at an angle of attack for which the wing drag coefficient is 0.015 and the parasite drag coefficient is 0.020. What is the speed in the dive from these altitudes: (a) 10,000 feet? (b) 20,000 feet? (c) 50,000 feet?
3. Why isn't it advisable to pull out of a dive abruptly?

CURVILINEAR (TURNING) FLIGHT

In rectilinear flight, the airplane was considered at any instant as flying at a steady speed with acceleration. In turning flight, there must be some acceleration to produce the necessary centrifugal force to keep the airplane in the turn.

Centrifugal Force

If a weight at the end of a rope is whirled around in a horizontal plane by the experimenter, he will find that there is a definite pull along the axis of the rope. This is the centrifugal force. It can be established that the centrifugal force (CF) can be calculated from the formula

$$CF = \frac{Wv^2}{gr}$$

where   CF = the centrifugal force, in pounds
        W = the weight of the object being swung
        v = the tangential velocity, in feet per second
        g = the gravitational constant
        r = the radius of the turn

The ratio v^2/r is the angular acceleration necessary to keep the mass (W/g) moving in a circle of radius r about the center O. Fig. 66 is graphical representation of an object turning about the point O.

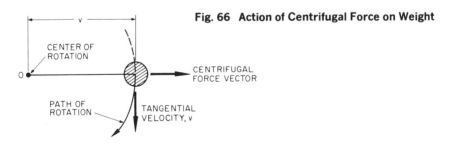

Fig. 66 Action of Centrifugal Force on Weight

Forces on the Airplane

There is no rope held by a post or by an experimenter to counteract the centrifugal force when an airplane makes a proper turn, but some

Forces on the Airplane (cont'd)

force component must hold the airplane in the turn. This component is obtained from the lift on the airplane when it is banked to the proper angle and makes the correct turn. In Fig. 67, the forces acting on the airplane (other than those perpendicular to the view shown) are:

- L = The lift perpendicular to the flight path and the plane of the wings, in pounds.
- W = The weight of the airplane, in pounds.
- CF = The centrifugal force, in pounds (acts at the center of gravity of each individual mass item of the airplane, but the net effect is the same as if it were acting through the center of gravity of the entire plane).

The airplane is banked at angle ϕ to the horizontal. Accordingly, the lift force, or vector, makes an angle ϕ with the vertical.

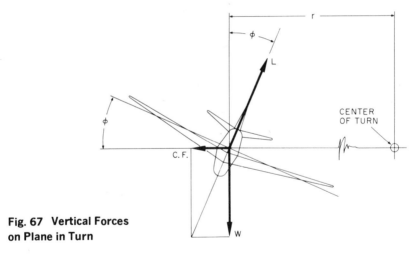

Fig. 67 Vertical Forces on Plane in Turn

Summation of Forces

The forces must be in equilibrium so that there is no outward motion of the airplane. It is convenient to use the gravitational axis as one of the coordinate axes, and a horizontal axis through the center of gravity for the other.

$$\Sigma V = 0 \text{ (Summation of all the forces in the vertical direction equals zero.)}$$
$$= L \cos \phi - W = 0$$

or

$$L \cos \phi = W$$

Summation of Forces (cont'd)

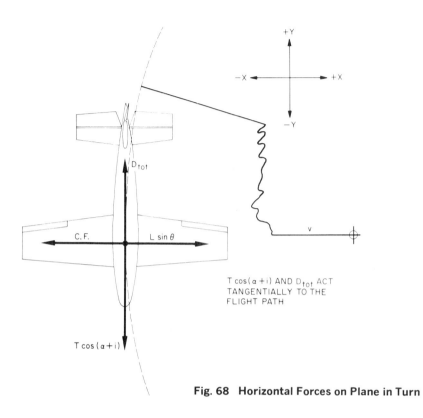

Fig. 68 Horizontal Forces on Plane in Turn

The contribution of the thrust component in the vertical direction can be assumed negligible for the same reasons as assumed when considering horizontal flight (sine values of small angles may be considered equal to zero.)

$\Sigma H = 0$ (Summation of all the forces in the horizontal direction equals zero.)
$$= L \sin \phi - CF = 0$$

or

$$L \sin \phi = CF$$

Dividing the first equation by the second,

$$\tan \phi = CF/W$$

But since

$$CF = Wv^2/gr$$

Summation of Forces (cont'd)

then substituting for CF it follows that

$$\tan \phi = v^2/gr$$

Considering the forces shown in the horizontal plane (Fig. 68),

$$\Sigma Y = 0$$
$$-T\cos(a + i) + D_{tot} = 0$$

For small angles, $\cos (a + i) = 1$, so that

$$D_{tot} = T$$

Likewise,

$$\Sigma X = 0$$
$$L \sin \phi - CF = 0$$

which is the equation obtained previously.

Analysis of Equations

It is apparent from the first equation that the lift in the turn must be greater than in rectilinear horizontal flight, for, in the latter, the lift (L) is equal to the weight (W) whereas in the turn, a component in the vertical direction must be great enough to balance the weight. Otherwise, the airplane would sink. To obtain the greater lift it is necessary:

1. That the airplane fly faster than in rectilinear (horizontal) flight at the same angle of attack.
2. If it flies faster, there must be greater drag since that force varies as the square of the speed, which is increased.
3. To obtain more thrust or power to accommodate greater speed (with its concomitant greater drag).

The second equation makes it apparent that the lift of the airplane must have a horizontal component to counteract or balance the centrifugal force. Otherwise, the airplane will skid in the direction of the centrifugal force.

From the observations made, it is also apparent that a turn cannot be made without skidding if the airplane is not banked appropriately, since there would be no horizontal component of the lift to offset the centrifugal force. Gentle turns might be possible without banking if the fuselage acted as a lift-producing surface in the turn.

Radius of Turn

The radius of turn can be found from the equations

$$L \sin \phi = CF$$

and

$$CF = \frac{Wv^2}{gr}$$

from which it follows that

$$r = \frac{Wv^2}{g L \sin \phi}$$

Since

$$L = (1/2) \rho C_L S v^2$$

then,

$$r = \frac{2W}{g \rho C_L S \sin \phi}$$

Several observations may be made regarding the radius of turn, considering one parameter at a time, with no change in the other parameters.

1. A larger radius is required for low values of C_L. Thus, for low values of α, the angle of attack, the radius of turn will be larger than for greater angles of attack.
2. The radius required to maintain a turn for a given angle of attack and angle of bank increases with altitude, since the air density (ρ) decreases.
3. The heavier the wing loading $\frac{W}{S}$, the greater the radius of turn required.
4. The smaller the angle of bank (ϕ), the greater the radius of turn required, since $\sin \phi$ approaches zero as ϕ approaches zero.

Example: An airplane with a gross weight of 8000 pounds is in a turn at 10,000 feet, with an angle of bank of 45 degrees. The angle of attack is 8 degrees. The airfoil characteristics are shown in the graph in Fig. 41. The parasite resistance coefficient referred to the wing area is 0.028. Determine the following for the angle of attack indicated:

(a) The speed in the turn. (Compare this with the speed in horizontal flight.)
(b) The lift on the airplane.
(c) The load factor on the airplane in the turn.

Radius of Turn (cont'd)

(d) The centrifugal force exerted on the airplane.
(e) The radius of turn.
(f) What is the drag on the turn? (Compare it with that in horizontal flight.)

Solution: The available data is: W = 8000 pounds, S = 500 square feet, α = 8., ϕ = 45°, C_{D_p} = 0.028. Airfoil properties are given in Fig. 41, from which C_D = 0.068 and C_L = 0.975 for the angle of attack of 8 degrees may be obtained. From the table for ICAO Standard Atmosphere (see page 8), the value of 0.00165 at 10,000 feet is indicated.

From the trigonometric tables, it will be found that tan 45° = 1, cos 45° = 0.7071, and sin 45° = 0.7071.

Since $L \cos \phi = W$ and $L = (1/2) \rho C_L S v^2$, combining the two and solving for v results in the formula

$$v = \sqrt{\frac{2W}{\rho C_L S \cos \phi}}$$

The lift of the airplane may be calculated from the formula

$$L = (1/2) \rho C_L S v^2$$

or

$$L = W/\cos \phi$$

The centrifugal force $CF = L \sin \phi$, since $L = W/\cos \phi$. Combining the two results in the formula

$$CF = \left(\frac{W}{\cos \phi}\right) \sin \phi = W \tan \phi$$

The radius of turn may be calculated from the formula

$$r = \frac{2W}{gS \rho C_L \sin \phi}$$

or, from the formula,

$$r = \frac{v^2}{g \tan \phi}$$

Radius of Turn (cont'd)

The drag on the airplane can be calculated from the standard drag equation or from the formula: $D = L/(L/D)$, where L is the lift in pounds, and the L/D is the ratio of lift to drag, $C_L/C_{D\,tot}$, of the airplane.

(a) The speed in horizontal flight is

$$v = \sqrt{\frac{2W}{\rho C_L S}} = \sqrt{\frac{2(8000)}{(0.001655)(0.975)(500)}}$$

$$= 148.6 \text{ feet per second}$$

The speed on the turn is

$$v = \sqrt{\frac{2W}{\rho C_L S \cos\phi}} = \sqrt{\frac{2(8000)}{(0.001655)(0.975)(500)(0.7071)}}$$

$$= 175 \text{ feet per second}$$

The speed on the turn is markedly higher.

(b) The lift on the airplane is

$$L = \frac{W}{\cos\phi} = \frac{8000}{\cos 45°} = \frac{8000}{0.7071} = 11{,}300 \text{ pounds}$$

c) The load factor would be determined by evaluating

$$n = L/W$$

Since

$$L = W/\cos\phi$$

it follows that

$$n = \frac{1}{\cos\phi} = \frac{1}{\cos 45°} = \frac{1}{0.7071} = 1.41$$

Radius of Turn (cont'd)

(d) The centrifugal force exerted on the airplane would be

CF = W tan ϕ = 8000 (tan 45°) = 8000 (1) = 8000 pounds

(e) The radius of the turn is found by evaluating

$$r = \frac{v^2}{g \tan \phi}$$

Since $g = 32.2$ and $\tan \phi = \tan 45° = 1$,

$$r = \frac{(175)^2}{32.2 \, (1)} = 950 \text{ feet}$$

(f) The $(L/D)_{tot}$ at 8° angle of attack is

$$(L/D)_{tot} = \frac{0.975}{(0.028 + 0.068)} = 10.15$$

The lift in the turn has been calculated above as being equal to 11,300 pounds; hence, the drag is

$$D = \frac{\text{Lift}}{L/D} = \frac{11{,}300}{10.15} = 1112 \text{ pounds}$$

This compares with a drag in horizontal flight of

$$\frac{8000}{10.15} = 788 \text{ pounds}$$

Review Questions

For the problems below, the following information is furnished: gross weight = 2500 pounds, wing area = 250 square feet, aspect ratio of the

CURVILINEAR (TURNING) FLIGHT   139

Review Questions (cont'd)

wing = 6. The lift and drag coefficients of the wing for three different angles of attack are as follows:

Angle of Attack	C_L	C_D
2°	0.50	0.025
4°	0.65	0.032
6°	0.80	0.046

The parasite drag coefficient of airplane, referred to the wing area, may be taken as 0.020. The airplane is in a coordinated turn at a 30-degree angle of bank.

Calculate:

(a) The speed in the turn for each angle of attack. (Compare these with those previously calculated for horizontal flight.)
(b) The centrifugal force exerted on the airplane.
(c) The radius of turn for each of the three angles of attack.

CONTROLS

It is necessary to have some means of maneuvering the airplane about its three axes. Such means—the controls—depend on aerodynamic forces.

The Flap

The lift of an airfoil can be changed radically by hinging to it a rear portion that may be deflected angularly. Figure 69 shows such an arrangement. The flap may be moved either downward or upward. By moving the flap down, the lift of the whole surface is increased in the upward direction; the converse occurs when the flap is moved upward. If a symmetrical airfoil is used, the lift increase for the flap up or down would be the same for the same angle of deflection.

Figure 70 shows the effect of the flap on the lift coefficient. The flap is called a *plain flap,* with a chord 30 percent of the airfoil chord, and was deflected in 5-degree increments from zero to 45 degrees. It will be noted that the new lift curves of C_L vs α are practically parallel. The increase in the maximum lift coefficient is of interest when the flap is used on the wing proper, but is of no interest when tail surfaces are considered (except that operation of the tail surfaces near the angle of maximum lift is generally not desired). There is an increase in the drag coefficient as well, but it is not considered significant.

The effect of the flap is to increase the lift coefficient of the surface with the flap down, as compared with that of the surface with the flap in the neutral position. This increase may be expressed as

$$\triangle C = (C_L) \text{ flap deflected} - (C_L) \text{ flap in neutral}$$

The principle of a lifting surface with a hinged flap is used for longitudinal, lateral, and directional control.

Longitudinal Control

The horizontal tail surfaces serve at least two functions: (1) to maintain trim (see *Longitudinal Stability*), and (2) to vary the trim so that the airplane can fly at a different angle of attack. The angle at which trim occurs is the angle of attack. At the angle of trim, the summation of the pitching moments about the center of gravity of the airplane is zero. (The

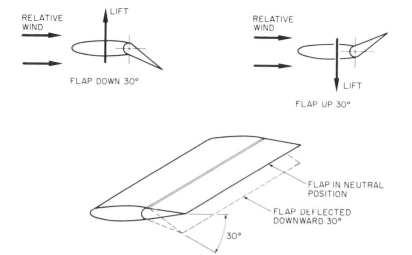

Fig. 69 Using a Flap as a Control Surface

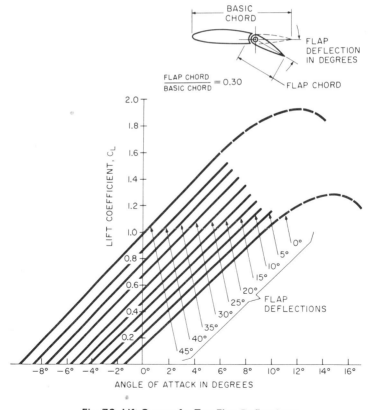

Fig. 70 Lift Curves for Ten Flap Deflections

Longitudinal Control (cont'd)

calculation of these pitching moments will be discussed under *Longitudinal Stability*.) Typical curves for the pitching moments are shown in Fig. 71.

The elevator is considered satisfactory if the angle of trim with an upward deflection of 30° of the elevator corresponds to the angle of attack at which the maximum lift coefficient of the wing occurs. The airplane is designed to trim with the elevator in the neutral position at the angle of attack corresponding to maximum speed. This angle of attack, at this speed, is at a relatively low lift coefficient. Therefore, not so much downward deflection as upward deflection of the elevator is required in flight. However, for takeoff purposes, airplanes employing the tail-wheel type of landing gear require a larger downward deflection of the elevator to get greater lift at the tail to place the airplane in takeoff position.

In Fig. 72, the pitching moment produced by the tail surfaces would be calculated by the equation

$$M_t = L_t C$$

A change in lift can be obtained by increasing the angle of attack of the horizontal tail surfaces, or by deflecting a flap portion (known as the elevator) with the front half remaining stationary. The deflecting elevator is more effective in changing the lift.

A typical planform of the horizontal tail surfaces is shown in Fig. 73. The horizontal tail surface areas vary in size from 10 to 15 percent of the wing area. The aspect ratio is generally low, from 3 to 5. The thickness ratio of the airfoil is from 6 to 9 percent. The surfaces are located from two to three times the length of the mean geometric chord behind the center of gravity of the airplane. Usually, the chord of the elevator is from 40 to 50 percent of the chord of the horizontal tail surfaces.

The stabilizer may be made adjustable so that an angular range of 2 to 5 degrees from neutral can be achieved.

The elevator is deflected downward to obtain an increase in lift upward. When this is done for an airplane having its horizontal tail surfaces aft of the center of gravity, the airplane will pitch down. If the horizontal tail surfaces are in front of the center of gravity, the airplane will pitch up. The motions of the airplane are the reverse when the elevator is deflected upward.

The elevator is deflected by a system linking the hinge of the elevator to a control column in the cockpit. If the pilot wishes to pitch the airplane down, he pushes forward on the control column. If he wishes to pitch the nose of the airplane up, he pulls back on the control column. (These movements are psychologically sound since the movements are instinctive.)

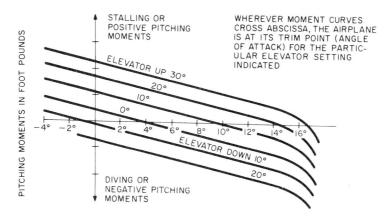

Fig. 71 Elevator Deflection Effect on Trim Point

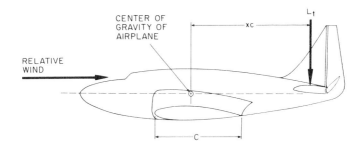

Fig. 72 Pitching Moment Caused by Elevator

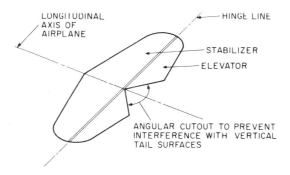

Fig. 73 Planform of Horizontal Tail Surfaces

Lateral Control — Ailerons

The airplane is moved about the longitudinal axis ("rolling") with *ailerons,* which are small flaps set in the trailing edge of the wing tips. One aileron is moved downward to increase the lift, while the other aileron simultaneously moves upward to reverse the lift at its tip. Ailerons so connected are known as "differential ailerons." The relative angular motions can be adjusted with linkage systems. The interconnection serves two purposes: the effect of increase in lift at one tip and decrease at the other augments the effect over that obtainable if only one aileron were used. The mechanical linkages are also simplified, since one aileron need not remain in neutral as the other is being deflected.

Figure 74 shows how the forces produced by the ailerons work. For the illustration shown, the expression for the rolling moment about the longitudinal axis is

$$L' = L_1 \frac{x}{2} + L_2 \frac{x}{2} = \frac{x}{2}(L_1 + L_2)$$

where L_1 and L_2 are the changes in the lift caused by the aileron flaps for the shaded area
L' represents the rolling moment

Calculations can be carried out for the rolling moment, but usually wind tunnel tests are used to determine the values experimentally.

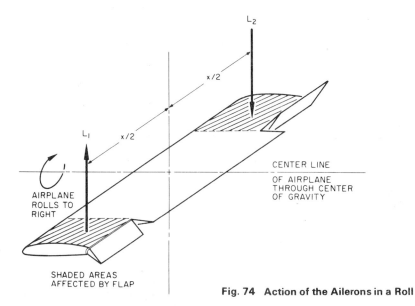

Fig. 74 Action of the Ailerons in a Roll

Lateral Control—Ailerons (cont'd)

When the ailerons are displaced, the airplane rolls about the longitudinal axis and continues to roll until the ailerons are returned to neutral. If the ailerons are just brought back to neutral, the airplane will not necessarily come back to even keel, but may continue to fly at a fixed angle of bank. To bring it to an even keel, the ailerons have to be displaced, to some degree, in the opposite direction. Of course, once motion comes into play, lateral stability effects are encountered—the return to an even keel will be achieved after a series of oscillations.

Variation with Angle of Attack

The differential aileron is not as effective at high angles of attack as it is at the higher speeds that are to be expected at low angles of attack. Typical curves are shown in Fig. 75.

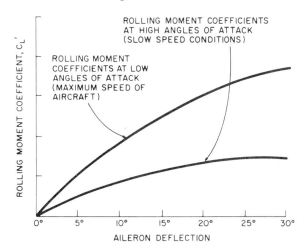

Fig. 75 Representative Rolling Moment Curves

The ailerons are even less effective when lift increase devices are used. Various devices have been used to increase the effectiveness of ailerons at high angles of attack of the wing. Since these are many, no attempt will be made to analyze the various methods available.

Adverse Yawing Moments

There is what might be called a side effect caused by the displacement of the ailerons. Due to the greater drag on the side of increased lift (as

Adverse Yawing Moments (cont'd)

when the aileron is deflected downward, as against the other side where the aileron is deflected upward), a yawing moment is introduced. This

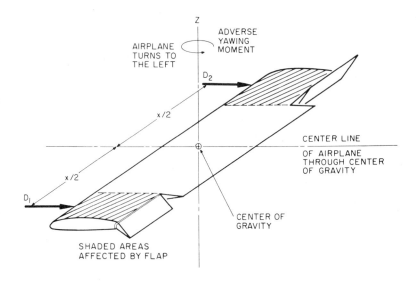

Fig. 76 Effect of the Ailerons on Yaw

yawing moment tends to force the airplane out of the turn for which the airplane is banked, as shown in Fig. 76. Hence, it is called an adverse yawing moment (see also *Turning Flight*).

Referring to Fig. 76, the yawing moment about the vertical axis is

$$N = -D_1 \frac{x}{2} + D_2 \frac{x}{2} = \frac{x}{2}(D_2 - D_1)$$

Since D_1 is greater than D_2, the yawing moment is negative. It tends to swing the airplane to the left while the ailerons are banked to the right for a right-hand turn.

The adverse yawing moment is greater for the airplane at high angles of attack than at low angles of attack, as illustrated in Fig. 77.

To maintain the airplane in a roll without turning from a rectilinear path, the vertical tail surfaces have to overcome merely the adverse yawing moments of the ailerons.

In the case of the coordinated turn (see *Turning Flight*), the vertical tail surfaces not only have to overcome the adverse yawing moments, but also must produce additional yawing moments to maintain the turn.

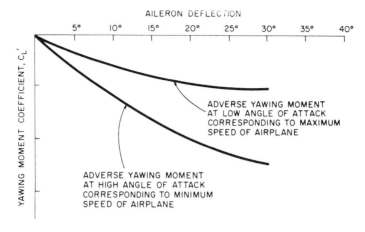

Fig. 77 Curves Showing Adverse Yawing Moments

Cockpit Controls

To roll or bank the airplane so that the right wing tip is down, the pilot moves the control column to the right. If he wants to bring the left wing tip down, he pushes the control column to the left. (These are instinctive motions and, therefore, easily learned.)

Directional Controls

The vertical tail surfaces serve not only to assure directional stability, but also to produce effective yawing moments to turn the airplane about its vertical axis. The disposition of areas of the vertical tail surface of an airplane is much the same as for the horizontal tail surfaces. In Fig. 78,

Fig. 78 The Vertical Tail Surfaces

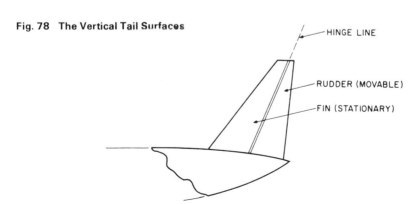

Directional Controls (cont'd)

the *fin* is the fixed portion, and the *rudder* the movable portion. The rudder usually rotates about a vertical axis. It works with the fin, just as the elevator works with the stabilizer.

The aerodynamic aspects of the vertical tail surfaces are much the same as those of the horizontal tail surfaces, except that the forces act along a different axis.

A turn to the right is produced by the yawing moment $L_t \cdot C$ as shown in Fig. 79. If the airplane were not banked, the airplane would skid (see *Turning Flight*). The rudder, in conjunction with the stabilizer, has to offset the adverse yawing moment produced by the ailerons. In addition, it must produce sufficient yawing moment to cause a turn.

Hinge Moments

All movable surfaces, such as ailerons, elevators, and rudders, have a moment about their axis of rotation due to the forces acting on them. For the sketch shown in Fig. 80, the moment about the axis of rotation, or hinge, would be $P_1 a$. This moment (plus any introduced by friction in the control system) is the moment that the pilot has to overcome when he uses the control column for banking or pitching the airplane, or when he uses the rudder pedals to operate the rudder.

Servo Control Tab

As the control surfaces—such as ailerons, elevators, and rudders—become larger, the forces to be exerted by the pilot also become larger. These movable surfaces can be moved more easily by operating a small flap, called a *tab*, set in the trailing edge of the large movable surface. The tab can be so designed that when the smaller tab is deflected, it will move the larger surface. When such a tab is operated from the cockpit, it is called a *servo control tab*. It may also be used to hold a movable surface in a certain deflected position.

Since the force on the tab is in a direction opposite to the total force on the larger deflected surface (see Fig. 81), the hinge moment about the hinge line of the larger surface would be

$$P_1 a - P_2 b$$

A tab doesn't need a very large surface area to help reduce the hinge moments, since not only the force on the tab, but also its moment arm from the hinge line of the main surface plays a role.

CONTROLS

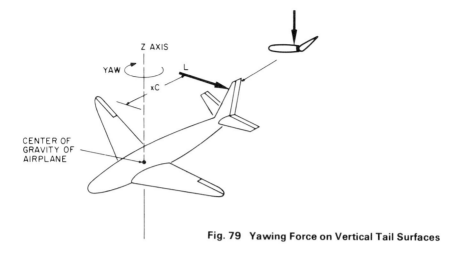

Fig. 79 Yawing Force on Vertical Tail Surfaces

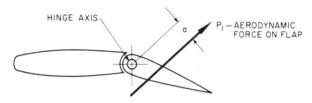

Fig. 80 Hinge Moment on a Movable Surface

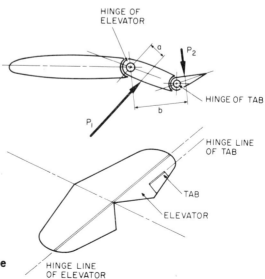

Fig. 81 Tab Used to Reduce the Hinge Moments

Review Questions

1. What is the function of a control surface? How does it usually operate?
2. How are rolling moments produced?
3. What adverse effect do ailerons have?
4. What kind of airfoils are normally employed for vertical and horizontal tail surfaces?
5. How is pitch of an airplane produced?
6. What are hinge moments?
7. What are servo controls?

STATIC STABILITY

Aircraft flight has other aspects than speed, glide, turn, and climb performance. What happens when a gust interferes with those inherent characteristics of the airplane that maintain it on course? What happens when the pilot momentarily deflects a particular control and then returns it to its original position? Will the airplane reach equilibrium again?

Equilibrium

A body is in a state of equilibrium when the sum of all the forces and the sum of all the moments caused by such forces are equal to zero. In this state, the airplane would remain in steady flight with no acceleration in any direction until disturbed—at which time it would not be in equilibrium. Its subsequent behavior would then become important, and will be explored by studying its *stability*.

Static Stability

The inherent ability of an object to return to its original status or condition of equilibrium (after being disturbed from that position or condition) is called *static stability*. If the object returns to its initial state of equilibrium after a disturbance, it is said to have positive static stability, or is *stable*. If it does not return to its initial position, or if it tends to deviate still further (usually in the direction of the force of the disturbance), then the object has negative static stability and is said to be *unstable*. If it assumes a new position of stability, it is said to be neutrally stable, or have *neutral stability*.

These states can be illustrated simply by considering a cone.

In Fig. 82a, if the cone resting on its base is tipped slightly either right or left, it will return to its original position once the force tipping it off its base is removed. This is an example of positive stability; the cone is stable.

In Fig. 82b, if the cone is held in a vertical position with its tip down, it will fall once the holding force is removed. It may be perfectly balanced in this position, but even a slight disturbance will upset it; and once upset, it will not return. This is an example of negative stability; the cone is unstable.

Static Stability (cont'd)

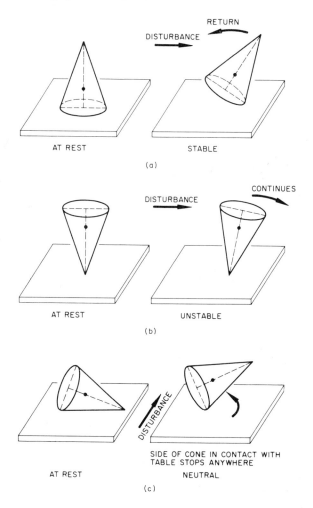

Fig. 82 Three Types of Static Stability

In Fig. 82c, the cone is lying on its side on a smooth, level surface. A slight push will make it roll to another position and remain there without any inherent tendency for it to return to its original position. The stability of the cone in this case is said to be neutral.

Since the airplane has three axes about which motion can take place, it follows that any one of the conditions of static stability could apply to the airplane in either pitching, yawing, or rolling—a total of nine possibilities. Yawing and rolling characteristics are interdependent in that rolling action is accompanied by yawing action, and vice versa.

Static Longitudinal Stability

In pitching, the static stability characteristics of an airplane can be calculated to a reasonable degree of accuracy. The effects of interference between fuselage, landing gear, power nacelles, as well as possible blockage effects of these components on tail surfaces, however, can only be surmised. Extensive wind tunnel testing on models is necessary to determine any such effects. If there are any interferences, design changes are made until the wind tunnel tests show that the desired characteristics have been obtained. In the final analysis, full-scale flight testing of a prototype model is required.

To be considered stable in flight, the airplane must have such inherent characteristics that when it is disturbed (from a state of steady flight at a given angle of attack), it will return to that angle of attack and continue its steady flight. If a disturbance should nose the airplane up momentarily, its inherent characteristics would automatically return it to its original state. It does not necessarily do this at once, but may oscillate. This behavior is related to dynamic stability, to be discussed later.

Pitching Moment Calculations

The pitching moments of an airplane must be calculated for each particular configuration, should it have more than one (due to deployment of lift increase devices, for example).

Figure 83 represents a fairly typical case. Instead of adhering to the conventional X-Z coordinates with the third quadrant for positive X and

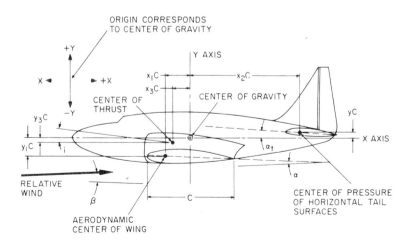

Fig. 83 Reference Dimensions for Calculation of Pitching Moments

Pitching Moment Calculations (cont'd)

Z ordinates, the more familiar X-Y axes will be used, and the first quadrant considered as having all positive coordinates. The center of gravity will be used as the origin of the coordinate system, since one condition for equilibrium is that the sum of the moments about the center of gravity be equal to zero at a particular angle of attack, known as the angle of trim.

Reference Dimensions

To simplify subsequent equations, all dimensions are expressed in terms of fractions or decimal parts of the mean aerodynamic chord. The necessary dimensions for the configuration shown are:

C = Length of the mean aerodynamic chord either in feet or inches. (All dimensions must, of course, be in the same units.) For preliminary calculations, the mean geometric chord may be used in place of the mean aerodynamic chord.

$x_1 C$ = The distance along the X direction of the aerodynamic center of the wing ahead of the Y-axis. Since it is a dimension to the left of the Y-axis, it would be negative.

$x_2 C$ = Distance between the assumed center of pressure, referred to the mean geometric chord, of the horizontal tail surfaces and the Y-axis. Since the dimension is to the right of the Y-axis, it would be positive.

$x_3 C$ = Distance from the center of thrust to the Y-axis. For a propeller-operated airplane, it would be at the intersection of the plane of rotation and the line of thrust.

$y_1 C$ = The distance of the wing aerodynamic center below the center of gravity. It is negative.

$y_2 C$ = The distance of the center of pressure of the horizontal tail surfaces above the center of gravity. It is positive.

$y_3 C$ = The distance the thrust is below the center of gravity. It is negative.

Forces Involved

The dimensions are shown in Fig. 83. The forces acting on the airplane should also be indicated on such a diagram, but the resultant sketch would be very cluttered. Hence, the force diagram is shown in Fig. 84, without the accompanying reference dimensions.

The forces acting on the airplane in steady flight are:

T = Thrust, of the power plant, in pounds.

Forces Involved (cont'd)

Fig. 84 Forces and Moments Acting on Plane

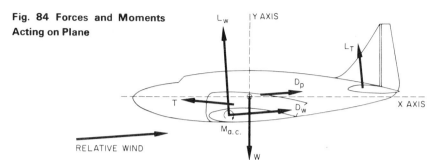

- L_w = Lift of the wing, in pounds: $L = qCLS$.
- D_w = Drag of the wing, in pounds: $D = qCDS$.
- D_p = Parasite drag of the airplane, in pounds. It includes all the drag of the airplane except that of the wing. For the sake of convenience, its line of action is assumed to be through the center of gravity. It is a reasonably accurate assumption that an airplane is more or less symmetrical about the longitudinal axis: $D_p = qC_{D_p}S$.
- L = Lift of the horizontal tail surfaces, in pounds: $L_t = qC_{L_t}S_t$.

Moments Involved

The pitching moments are to be calculated.

- $M_{a.c.}$ = Wing moment, in foot pounds (if dimensions are in feet), about the aerodynamic center of the wing to which the airfoil coefficients C_L and C_D were referred. (See *Airfoil Characteristics*.) The relevant equation is

$$M = qC_{M_{a.c.}} CS$$

- $M_{c.g.}$ = Pitching moment of the airplane about its center of gravity.

Angles

The diagram shows designations for a number of angles.

- β = Angle of flight with reference to the horizon. However, the stability of the airplane is unaffected by the flight path, whether it be climbing, gliding, or horizontal.
- α = Angle of attack. This is the angle included between the line of flight and the mean aerodynamic chord of the wing.

Angles (cont'd)

i = Angle of incidence. If the thrust line is inclined to the chosen longitudinal axis of the airplane, the angle i is indicated. The wing and tail surfaces may have angles of incidence, too, but these have not been considered in this derivation since neglecting them does not invalidate the conclusions to be drawn later.

α_t = Angle of attack of horizontal tail surfaces. This angle will be different from that of the wing, but for preliminary calculations, the angle of attack may be assumed the same.

Speeds Involved

Only one speed need be considered—that speed in steady flight at which the disturbance took place.

Pitching Moment Equation

With all this information, one may now write a pitching moment equation referred to the center of gravity. Clockwise moments are considered positive, and counterclockwise moments are negative. Each factor affecting the moments will be considered separately, and then all the quantities summed up for the final equation.

1. Due to $C_{M_{a.c.}}$ $qSCC_{M_{a.c.}}$ — a moment can be replaced by a couple. It can then be proved that the value of a couple is the same about any point in the same plane.
2. Due to wing lift: $+ qSC_L y_1 C \sin \alpha$ for the horizontal component;
     $+ qSC_L x_1 C \cos \alpha$ for the vertical component.
3. Due to wing drag: $- qSC_L y_1 C \cos \alpha$ for the horizontal component;
     $+ qSC_L x_1 \, C \sin \alpha$ for the vertical component.
4. Due to tail lift: $- qS_t C_{L_t} x_2 C \cos \alpha$ for the vertical component;
     $+ qS_t C_{L_t} y_2 C \sin \alpha$ for the horizontal component.

Pitching Moment Equation (cont'd)

5. Due to parasite drag: $+ D$ (0 for both components, since it is being assumed that the parasite drag acts through the center of gravity)
6. Due to thrust: $- T y_3 C \cos(i + \alpha)$ for the horizontal component; $+ T x_3 C \cos(i + \alpha)$ for the vertical component.
7. Due to weight: $+ W$ (0 since moments are taken about the center of gravity through which the weight acts).

Before summing up the terms above, some simplifications can be made: sine values of small angles are considered to be equal to zero, and cosine values of small angles equal to 1; the contribution of the thrust will be considered separately, since, with the assumptions for small angles made above, the quantity will remain constant.

Collecting all of the quantities listed for 1, 2, 3, and 4, and noting that qSC is a term common to all, the pitching moments about the center of gravity become

$$M_{c.g.} = qSCC_{M_{c.g.}} = qSC \left[C_{M_{a.c.}} + x_1 C_L - y_1 C_D - x_2 C_{L_t} (S_t/S) \right]$$

Dividing both sides of the equation by qSC,

$$\frac{M_{c.g.}}{qSC} = C_{M_{c.g.}} = C_{M_{a.c.}} + x_1 C_L - y_1 C_D - x_2 C_{L_t} (S_t/S)$$

where S is the wing area

This equation is convenient, since it does not entail as much calculation as when the multiplying factor (qSC) is included.

Tabular Calculations

The calculations can be carried out in tabular form, as shown, which is most convenient, since repetitive calculations are required.

Example: To illustrate the use of the table for calculating terms in the moment equation, calculations will be carried out for one angle of attack

Tabular Calculations (cont'd)

CALCULATING MOMENT EQUATION TERMS*													
1	2	3	4	5	6	7	8	9	10	11	12	13
a_w^o	C_L	x_1	$C_L x_1$	C_D	$y_1 C_D$	4−6	$7+C_{M_{a.c.}}$	a_t^o	C_{L_t}	$y_2 C_{L_t}$	$11(S_t/S)$	8−12
−6												
−4												
−2												
0												
2												
4												
6												
etc.												

*Col. 1: List the angles of attack of the wing as shown in the table.
Cols. 2 & 5: From the aerodynamic characteristics of the wing, list the values of C_L and C_D corresponding to the angles of attack indicated in Col. 1.
Col. 3: The value of x_1 is constant since this is a function of the structural configuration of the airplane.
Col. 4: Multiply the values of C_L (from Col. 2) by x_1.
Col. 5: See above.
Col. 6: The value of y_1 is constant for the same reason that x_1 is. Multiply values of C_D (from Col. 5) by y_1.
Col. 7: From the value of $x_1 C_L$ obtained in Col. 4, subtract the value of $y_1 C_D$ obtained in Col. 6.
Col. 8: To the values in Col. 7, add the value of $C_{M_{a.c.}}$. The value of $C_{M_{a.c.}}$ is constant.
Col. 9: List the *angle of attack of the horizontal tail surfaces* when the wing is at the *angles of attack of the wing* listed in Col. 1.
Col. 10: From the aerodynamic characteristics of the horizontal tail surfaces, list the lift coefficient, C_{L_t}, corresponding to the values of the angles of attack listed in Col. 9.
Col. 11: Multiply the values of C_{L_t} (from Col. 10) by y_2 which is a constant for the airplane.
Col. 12: Multiply the quantities in Col. 11 by the ratio of S_t/S to obtain a value referred to the wing area S.
Col. 13: From the values calculated in Col. 8, subtract the values calculated in Col. 12. The values thus obtained are designated $C_{M_{c.g.}}$.

for an airplane with a wing whose aerodynamic characteristics are those shown in Fig. 41:

Wing area, S	500 sq ft
Mean geometric chord	10' - 0"
Configuration as shown in Fig. 83	
$x_2 C$	25' - 0"
$x_1 C$	1' - 0"
$y_1 C$	1' - 6"
Horizontal tail surface area, S_t	75 sq ft

Choosing the 8-degree angle of attack

$$C_L = 0.975 \qquad C_D = 0.068 \qquad C_{M_a} = -0.09$$

The lift coefficient of the horizontal tail surface is $0.02 = C_{L_t}$

Tabular Calculations (cont'd)

Solution: The applicable formula is

$$C_{M_{c.g.}} = C_{M_{a.c.}} + x_1 C_L - y_1 C_D - x_2 C_{L_t}(S_t/S)$$

Since $x_1 C = 1'\text{-}0''$ and $C = 10'\text{-}00''$

$$x_1 = 0.10$$

Since $x_2 C = 25'\text{-}0''$ and $C = 10'\text{-}0''$

$$x_2 = 2.5$$

Since $y_1 C = 1'\text{-}6''$ and $C = 10'\text{-}0''$

$$y_1 = 0.15$$

Moreover,

$$S_t/S = 75/500 = 0.15$$

Referring to the table, enter on the line for an attack angle of $+8°$ the values:

In column 2: $C_L = 0.975$.
" " 3: $x_1 = 0.100$ (will be constant for all angles).
" " 4: $C_L x_1 = (0.975)(0.10) = 0.0975$.
" " 5: $C_D = 0.068$.
" " 6: $y_1 C_D = (0.15)(0.068) = 0.0102$.
" " 7: Subtract the value in column 6 from that in column 4, or $(0.097) - (0.0102) = 0.0873$.
" " 8: Add $C_{M_{a.c.}}$ to the value in column 7, or $(0.0873) + (-0.0900) = -0.0027$.
" " 9: Indicate the angle of attack of the horizontal tail surfaces. This may be somewhat different than the wing angle of attack.
" " 10: Enter the lift coefficient for the horizontal tail surfaces corresponding to the angle of attack listed in column 9. The value used here, for illustrative purposes, is $C_{L_t} = 0.02$
" " 11: $x_2 C_{L_t} = 2.5(0.02) = 0.05$.

Tabular Calculations (cont'd)

In column 12: Multiply the quantity in column 11 by $S_t/S = 0.15$, or $(0.05)(0.15) = 0.0075$.

" " 13: Subtract the quantity in column 12 from that in column 8, or $(-0.027) - (0.075) = 0.0102$.

This is a point to be entered on a graph for C_{M_t} such as is shown in Fig. 85 (which is a graph for another airplane).

These calculations are carried out in tabular form and have to be derived for each configuration of the particular airplane design under consideration.

The same sort of table could be set up for the equation for which the assumptions for $\cos \alpha \approx 1$ and $\sin \alpha \approx 0$ had not been made.

The data obtained in columns 6, 7, 8, 11, and 13 represent the quantities expressed in the derived formula for $C_{M\ c.g.}$ with the last column giving the final calculated value for the corresponding angle of attack.

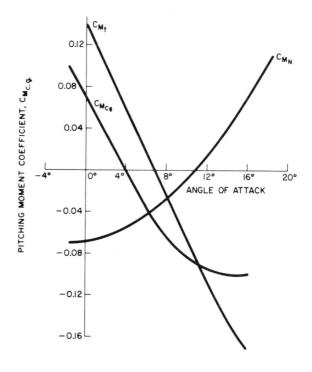

Fig. 85 Pitching Moment Coefficients of Wing, Tail, and Plane about Center of Gravity

Analysis

The quantities $C_{M_{a.c.}} + x_1 C_L - y_1 C_D$ constitute the contribution to the total made by the forces and moments on the wing, and may be designated C_{M_w}. The quantity $-(x_2 C_{L_t} S)/S$ constitutes the contribution to the total made by the tail surfaces and may be designated C_{M_t}. The longer equation could then be rewritten

$$C_{M_{c.g.}} = C_{M_w} + C_{M_t}$$

Values for $C_{M_{c.g.}}$ are to be found in column 13; values for C_{M_w} in column 8; and values for C_{M_t} in column 12. All of these values can be plotted against the corresponding values in column 1. Figure 85 represents these curves.

Examination of the $M_{c.g.}$ curve shows that the airplane for which it was calculated was stable since:

1. The value of $C_{M_{c.g.}}$ is zero at an angle of attack of 4 degrees. This angle is known as the angle of trim. One of the conditions for equilibrium, in addition to the fact that the summation of the forces acting on the airplane must equal zero, is that the summation of the moments must equal zero at a particular angle of attack, known as the angle of trim. Even though the quantity shown is $C_{M_{c.g.}}$, it need only be multiplied by a constant (qSC) to obtain the pitching moment, $M_{c.g.}$.

2. The slope of the curve is such that, as the angle of attack is increased about 4 degrees, the pitching moment coefficient becomes negatively larger. Thus, the pitching moment will force the nose of the airplane down or, in other words, will counteract the effect of the disturbance forcing the nose of the airplane up.

    Similarly, if the disturbing force decreased the angle of attack below 4 degrees, the $C_{M_{c.g.}}$ becomes larger in the positive direction. In other words, the stalling moments become larger, counteracting the effect of the disturbing force to force the nose of the airplane down.

Degree of Stability

It is important to know *how* stable an airplane is. Are its inherent characteristics, that is, the counteracting pitching moments, such that

Degree of Stability (cont'd)

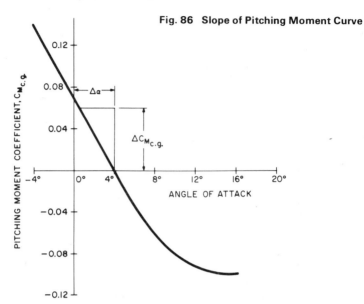

Fig. 86 Slope of Pitching Moment Curve

the airplane returns slowly or aggressively to its angle of trim? This response characteristic is known as the *degree of stability*.

The degree of stability is measured by the slope of the curve. Referring to Fig. 86, the slope is measured by the tangent drawn to the curve. At or near the angle of trim, the curve is a straight line, so the tangent coincides with the curve.

$$\text{The tangent} = -\frac{\Delta C_{M_{c.g.}}}{\Delta a}$$

The negative sign indicates that the curve slopes upward to the left. The negative slope is necessary so that the conditions discussed in (1) and (2) can apply—and which must apply if the airplane is to be stable.

Recapitulating, the necessary conditions for an airplane to be longitudinally stable are:

1. That the pitching moments total zero at an angle of trim within the range of operating angles of attack.
2. That the slope of the pitching moment or pitching moment coefficients, versus the angle of attack, be negative. The greater the negative slope, the greater the stability of the airplane.

The curve for C_{M_w}, representing the wing pitching moments, is worth examining. Note that, while there is an angle at which the moments equal

Degree of Stability (cont'd)

zero, the slope of the curve is such that if a disturbing force momentarily increased the angle of attack, it would tend to continue increasing. Hence, the wing alone would be unstable.

Effect of Thrust

Earlier, it was suggested that the factor involving thrust in the pitching moment equation could be neglected for the time being. It should be noted that the quantity Ty_3C is a constant, and since the cosine of the angle is close to 1, changes in angle of attack (especially in the region where trim is desired) have practically no effect. The pitching moment curve would be displaced by a constant amount (see Fig. 87).

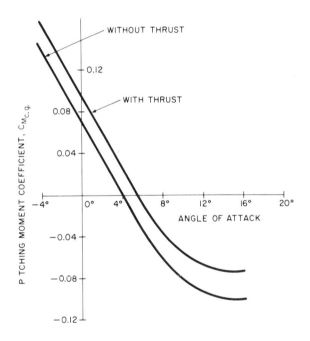

Fig. 87 Thrust Effect on Pitching Moment Curve

The thrust force, for which the moment equation was derived, caused a positive pitching moment. Therefore, the moment curve would be moved to the right. The angle of trim would, therefore, be greater with thrust than without it. Should the engine quit, the trim angle would be lower.

Effect of Thrust (cont'd)

It is generally desirable to have the thrust line passing close to or through the center of gravity so that the power-on and power-off conditions do not produce too great a change in the location of the angle of trim.

There may be changes of airflow over the tail surfaces when the power is on. These changes may change slightly the pitching moments produced by the horizontal tail surfaces.

Neutral Stability

In this case, the airplane has no one angle of trim, but many. This situation is illustrated in Fig. 88. The neutral stability zone is that between A and B.

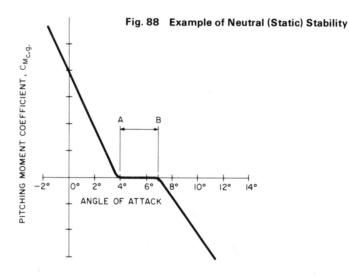

Fig. 88 Example of Neutral (Static) Stability

The condition shown is not one that can be detected by straightforward calculations, because it may be caused by interference with the proper airflow over the tail surfaces and wing. It can be detected in wind tunnel tests on scale models of the aircraft. If found there, the design would be corrected before a full-scale prototype was built.

Another variation of neutral stability is one that causes an airplane to "hunt" for its angle of trim since there are two or three possible angles of trim rather close together, as shown in Fig. 89.

Neutral Stability (cont'd)

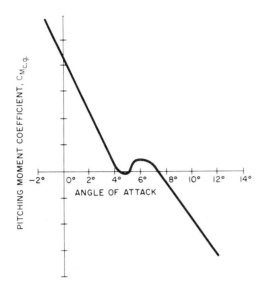

Fig. 89 This Airplane Has Neutral Stability and Three Angles on Trim

Review Questions

1. Define static stability.
2. Illustrate the various states of stability by using a ball instead of a cone.
3. How is the degree of longitudinal stability determined?
4. With reference to Fig. 85, what may be said about:
    (a) The stability of an airplane having no tail surfaces?
    (b) The role played by the horizontal tail surfaces in obtaining longitudinal stability?
5. If any, what is the effect of thrust on stability?

DYNAMIC STABILITY

Definition

The behavior of the airplane following a disturbance from its state of equilibrium depends on, and indicates, its *dynamic stability* characteristics.

No attempt will be made to discuss calculations that might be used to investigate the dynamic stability characteristics of an airplane. They are complicated—requiring higher mathematics and the use of a computer. Such calculations are seldom made for small airplanes of normal configuration. If unusual phenomena occur, they are checked by flight testing.

The various types of dynamic responses may be graphically represented to show angular displacements of the airplane plotted against time. (All the following discussions apply to the longitudinal stability of the airplane.)

Unstable Condition

A statically unstable airplane, once displaced, would continue to depart from the quilibrium or trim point. (It would show no oscillations about the spanwise or Y-axis.) Such a state is said to be *divergent*. This situation is illustrated in Fig. 90. A statically unstable airplane is *dynamically unstable*.

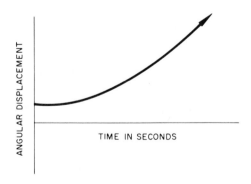

Fig. 90 Divergent or Negative Stability (Static Instability)

Unstable Condition (cont'd)

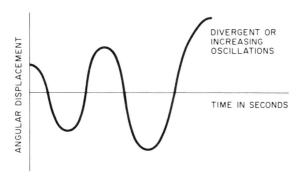

Fig. 91 Negative Dynamic Stability

A statically stable airplane may be dynamically unstable if the oscillations caused by a disturbance tended to increase with time, as shown in Fig. 91.

Neutral Dynamic Stability

An airplane that has neutral static stability will have *neutral dynamic stability* (as shown in Fig. 92).

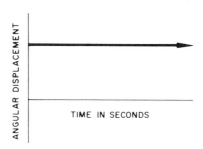

Fig. 92 Curve of Neutral Dynamic Stability

An airplane with positive static stability may have neutral dynamic stability (as shown in Fig. 93).
The oscillations neither increase nor decrease with time.

Neutral Dynamic Stability (cont'd)

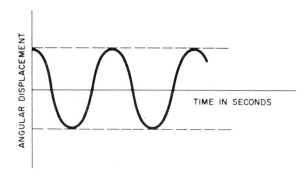

Fig. 93 Also a Case of Neutral Dynamic Stability

Positive Dynamic Stability

If, after the initial disturbance, the airplane returns to its original angle of trim in one motion, the airplane is dynamically stable. The condition is said to be *aperiodic* or *deadbeat damping*, as illustrated in Fig. 94.

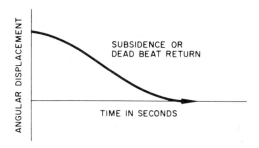

Fig. 94 Positive Dynamic Stability

If the oscillations continue but decrease with time, the airplane is also dynamically stable. The time history of these oscillations is shown in Fig. 95. The oscillations are said to be *damped*.

Positive Dynamic Stability (cont'd)

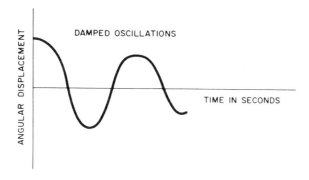

Fig. 95 Another Case of Positive Dynamic Stability

Review Questions

1. What is meant by dynamic stability? How is it measured?
2. Indicate the difference between positive static and positive dynamic stability.
3. Draw representative curves of:
   (a) Neutral dynamic stability.
   (b) Negative dynamic stability.
   (c) Positive dynamic stability.

LATERAL AND DIRECTIONAL STABILITY

The discussion on static and dynamic stability is equally applicable to lateral and directional stability. The material on controls should be considered with this material.

Rolling Moments

Rolling moments are brought into play in lateral stability. The pertinent equation is

$$L' = (1/2)\rho v^2 S b C_{L'} = q b S C_{L'}$$

where L' = rolling moment in inch-pounds, or foot-pounds, depending upon the units used for b, the span length. (The prime is used to distinguish the L from the notation used for lift.)
q = $(1/2)\rho v^2$, in pounds per square foot
v = the forward velocity of the airplane, in feet per second
S = wing area, in square feet
b = span of the wing, usually expressed in feet, whereupon the equation for the rolling moment will result in foot-pounds
C_L' = the rolling moment coefficient of the airplane. It is non dimensional and should not be confused with C_L, the lift coefficient.

The rolling moment is positive when the airplane rolls to the right, as viewed from the pilot's seat. These rolling moments are produced by the ailerons. (See *Controls*.)

Static Lateral Stability

The airplane is statically stable when it returns to its original lateral attitude after being disturbed. It is neutrally stable if it tends to stay in its new position when placed there by an initial disturbance. It is statically unstable if the airplane continues to roll.

Dynamic Lateral Stability

Applying the same definitions to the various types of dynamic stability as for longitudinal stability, the airplane is dynamically stable laterally

LATERAL AND DIRECTIONAL STABILITY

Dynamic Lateral Stability (cont'd)

if it returns in one motion to its original attitude after being disturbed (made to roll), or if the oscillations damp out gradually (each successive rolling oscillation becomes less, and the oscillations cease after a reasonable time.) The airplane is dynamically unstable if the rolling oscillations increase.

Use of Dihedral

Lateral stability may be obtained by providing a dihedral angle to the wing, as illustrated in Fig. 20. The angle of dihedral is the angle included between the horizontal (parallel to the Y-axis) and the span line of the wing. This angle may vary from 3 to 8 degrees, depending on how much of the wing is swept upward.

Yawing Moments

The static directional stability of an airplane is connected with its yawing characteristics. The airplane turns or yaws about its vertical or Z-axis. Positive yaw occurs when the nose of the airplane turns to the right. The yawing moment equation is expressed

$$N = (1/2) \rho v^2 SbC_n = qSbC_n$$

where N = yawing moment, in inch-pounds or foot-pounds, depending upon the units used for b, the span length
C_n = the non dimensional yawing moment coefficient based upon the wing area,
b = span of the wing, usually expressed in feet, whereupon the equation for the rolling moment will result in foot-pounds
v = the forward velocity of the airplane, in feet per second
q = $(1/2) \rho v^2$, in pounds per square foot
S = wing area, in square feet

It is convenient to refer the coefficient to the wing area since it is the main element of an airplane and the one on which performance characteristics depend.

Experimental methods, such as wind tunnel tests, are used to determine yawing moments of the aircraft. In general, it is desirable to have more effective fin area behind the center of gravity than in front of it. (All elements of the airplane such as the wing, fuselage, tail surfaces, and even the landing gear offer surfaces upon which the air impinges—they become part of the fin area.)

Use of Sweepback

Wing sweepback is used to obtain desirable directional stability characteristics. The sweepback or sweep angle is measured between the line perpendicular to the plane of symmetry and the line established by the aerodynamic centers of the wing, as shown in Figs. 18 and 19.

Review questions

1. What are the geometric parameters on which the rolling moment equation is based? The yawing moment equation?
2. Define positive static lateral stability.
3. Define positive dynamic lateral stability.
4. Define positive static directional stability.
5. Define positive dynamic directional stability.

Index

Index

Aerodynamic center of airfoil, 66–67
Aerodynamic data, presentation of, 50–53
Aerodynamic forces, 43–45
Aerodynamic similarity, 81–82
　Reynolds number, 80–81
Aerodynamic twist of wing, 63
Ailerons, 144–147
　adverse yawing moments, 145–147
　cockpit controls, 147
　effectiveness with variation in angles of attack, 145
Air
　Boyle's law, 4
　Charles' law, 4
　combined Charles' and Boyle's laws, 4–6
　composition, 3
　density, 3
　　effect on speed, 88
　effect of pressure and temperature, 4–6
　modulus of elasticity, 7
　pressures, 5
　properties of, 3–7
　temperature, 3–4
　viscosity, 6
Airfoil
　aerodynamic center, 66–67
　aerodynamic characteristics, 43–53
　aerodynamics, 54–69
　angle of attack, 43
　camber, 22–23
　center of pressure, 63–65
　chord line, 21–22
　coordinates, 24–27
　description, 21
　flow around, 12–14
　geometry, 21–27

Airfoil (*continued*)
　laminar flow, 14
　leading edge radius, 23
　lift over drag ratio, 63
　moment coefficient, 65–66
　pressure distribution, 43–45
　section characteristics, 67–68
　thickness, 23–24
　trailing edge, 23
Airflow, 9–17
　around an airfoil, 12–14
　Bernouilli's theorem, 11–12
　boundary layer, 15
　equation of continuity, 9–10
　laminar, 14
　Mach number, 17
　streamlines, 10
　stream tubes, 10–11
　supersonic, 16–17
　turbulence, 15
Angle(s)
　of attack
　　of airfoil, 43
　　variation, 57–59
　calculation of, 39–40
　of climb, 110–112
　dihedral, 38
　　use of, 171
　effect of thrust on trim, 163
　of glide, 60
　of incidence of wing, 38
　of motion of airplane, 85
　sweepback, use of, 172
　of trim, 140, 161
　of wing, 36
Aspect ratio
　corrections for drag coefficients, 60–63
　effect on angle of attack, 57–59

175

INDEX

Atmosphere
  definition of standard, 7–8
  ICAO standard, 8
Axes
  body, 41–42
  reference, 41–42
  wing, 42

Bernouilli's theorem, 11–12
Boundary layer, 15
Boyle's law, 4
Burble point, 55

Camber
  of airfoil, 22–23
  change by means of flaps, 74
Centrifugal force in turning flight, 131
Charles's law, 4
Chord
  airfoil, 21–22
  mean geometric, 28–35
  of wing, 28
Climbing flight, 108–117
Climb
  forces on airplane, 108–109
  power required, 112–116
  rate of, 112
  speed along path, 109–110
Composition of air. *See* Air
Controls, 140–150
  ailerons, 146
  elevator, 142
  the flap, 140–141
  hinge moments, 148
  longitudinal, 140–143
  servo tab, 148–149
Coordinates, airfoil, 24–27
Curvilinear flight, 131–139
  See also Turning flight

Density of air. *See* Air, density
Dihedral
  use of, 171
  wing, 38
Directional controls, 147–149
Directional stability, 170–172
  use of sweepback, 172
  yawing moments, 171
  See also Yawing
Dive, 127–130
  limiting speed, 128
  load factor in pull-out, 129
  pulling out of, 128–130

Diving moments (in pitching), 85
Drag
  aspect ratio corrections, 60–63
  coefficient, 59–60
  equation, 49
  forces, 47–48
  induced, 60–62
  interference, 71–72
  parasite, 70–71
  profile, 60
Dynamic stability, 166–169

Equation of continuity, 9–10
Elevator
  control linkage, 142
  use in longitudinal control, 142
  See also Horizontal tail surfaces; Stability, longitudinal
Experimental methods, 78–83

Flap
  as a control device, 140–141
  hinge moments, 148
  maximum lift coefficient, 74–77
  types, 74–75
Forces
  aerodynamic, 43–45
  on airplane in climb, 108–109
  centrifugal, 131
  in dive, 127–128
  in horizontal flight, 86–90
  lift and drag, 47–48
  resultant, 45–46
  in turning flight, 131–134
Fuselage, 1

Gliding flight, 118–126
Gust
  change in lift coefficient, 103–104
  effect of, 102
  effect on lift, 104
  load factor, 104–107

Horizontal flight
  horizontal forces, 92–93
  power required, 95–98
  variation of thrust, 93–95
  vertical forces, 86–88
Horizontal tail surfaces
  function of, 140–143
  planforms, 142–143

INDEX

ICAO Standard Atmosphere Table, 8

Laminar flow, 14
Landing, 122–125
  radius, 121–122
Landing gear, 1–2
Lateral control. *See* Ailerons
Lateral stability
  adverse yawing moments, 145–146
  dynamic, 170–171
  rolling moments, 170
  static, 170
  use of dihedral, 171
Leading edge radius of airfoil, 23
Lift
  burble point, 55
  coefficient, 54–56
  over drag ratio, 69
  effect of flaps, 74–77
  effect of gust, 104
  equation, 49
  forces, 47–48
  increase devices, 73–77
  maximum coefficient, 56
  slope of lift curve, 56–57
  variation
    with angle of attack, 49
    of speed with coefficient, 88–89
Load factor
  in dive, 129
  due to a gust, 104–107
Longitudinal control, 140–143
  See also Stability, longitudinal; Horizontal tail surfaces

Mach number, 17
Moment coefficient of airfoil, 65–66
Motion, 18–20
  of airplane, 84–85
  angles, of airplane, 85
  curvilinear, 84–85
  diving (in pitching), 85
  laws of, 18–19
  linear, 84
  pitching, 85
  relative, 19
  rolling, 85
  stalling (in pitching), 85
  yawing, 85

Pitching moment calculations, 153–160
Planforms
  horizontal tail surfaces, 143

Planforms (*continued*)
  vertical tail surfaces, 147
  wing, 35
Power
  coefficient, 100
  required in horizontal flight, 95–98
  variation with altitude in horizontal flight, 98–99
Power plant, 2
Pressure
  distribution over airfoil, 43–45
  dynamic, 49
Properties of air, 3–17

Radius of turn, 135–138
Rectilinear flight, 86–101
Reynold's number, 80–81
Roll, 85
Rudder. *See* Vertical tail surfaces; Yawing; Directional stability

Sound, velocity of. *See* Mach number
Span of wing, 28
Speed
  along glide path, 122
  in climb, 109–110
  criterion for landing, 73
  limiting, in dive, 128
  Mach number, 17
  regimes, 19–20
  relative, 19
  sinking, 120–121
  of sound, 17
  stalling, of wing, 80
  variation
    with air density, 88
    with altitude, 90
    with gross weight, 88
    with lift coefficient, 88
    with wing loading, 88
Stability
  degree of static longitudinal, 161–163
  directional, 170–172
  dynamic, 166–169
    neutral, 167–168
    positive, 168–169
    unstable, 166–167
  effect of thrust, 163
  lateral, 170–172
  longitudinal, 153
  static, 151–152
    "hunting", 164
    negative or unstable, 151
    neutral, 164

Stalling
    in pitching, 85
    of wing, 80
Standard Atmosphere Table, 8
Static stability, 151–165
Streamlines, 10
Stream tubes, 10–11
Supersonic airflow, 16–17
Sweepback, wing, 36–37
    use of, 172

Tail surfaces, 1
    See also Controls; Vertical tail surfaces; Horizontal tail surfaces
Thickness of airfoil, 23–24
Thrust
    effect on glide, 120
    effect on trim angle, 163
    variation in horizontal flight, 93–95
Trailing edge of airfoil, 23
Turning flight, 131–139
    centrifugal force, 131
    comparisons with horizontal flight, 134
    radius of turn, 135

Velocity, relative, 19
    See also Speed

Vertical tail surfaces, 147–148
Viscosity of air, 6

Wind, relative, 19
Wind tunnel, 78–80
    measuring system, 78–79
    method of operation, 78–80
    models for testing, 80
Wing
    angles, 36
    angle of incidence, 38
    aerodynamic twist, 63
    area, 39
    aspect ratio, 39–40
    center of pressure, 63–65
    chord, 28
    description, 1
    dihedral, 38
    geometry, 28–40
    loading effect on speed, 88
    planforms, 35
    root, 28
    span, 28
    sweepback, 36–37

Yawing, 85
    effect of ailerons, 145–147
    adverse moments, 145